17·99

Engineering GNVQ: Intermediate

Second edition

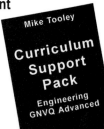

Engineering GNVQ: Intermediate

Second edition

Mike Tooley
Director of Learning Technology
Brooklands College of Further and Higher Education

Newnes

OXFORD   AUCKLAND   BOSTON   JOHANNESBURG   MELBOURNE   NEW DELHI

Newnes
An imprint of Butterworth-Heinemann
Linacre House, Jordan Hill, Oxford OX2 8DP
225 Wildwood Avenue, Woburn, MA 01801-2041
A division of Reed Educational and Professional Publishing Ltd

A member of the Reed Elsevier plc group

First published 1996
Second edition 2000
Reprinted 2001

British Library Cataloguing in Publication Data
A catalogue record for this book is available from the British Library.

ISBN 0 7506 4756 6

Printed and bound in Great Britain

For more information on all Butterworth-Heinemann publications please visit our website at www.bh.com

Success through qualifications

Edexcel recommends this book as suitable for its Intermediate GNVQ Engineering programme.

FOR EVERY TITLE THAT WE PUBLISH, BUTTERWORTH-HEINEMANN
WILL PAY FOR BTCV TO PLANT AND CARE FOR A TREE.

Contents

Preface

Welcome to the challenging and exciting world of engineering! This book is designed to help get you through the core units of the Intermediate General National Vocational Qualification (GNVQ) in Engineering. It contains all of the material that makes the essential underpinning knowledge required of a student who wishes to pursue a career in any branch of engineering.

This book was originally written by a team of Further and Higher Education Lecturers. They each brought with them their own specialist knowledge coupled with a wealth of practical teaching experience. This new edition has been designed to cover the core units of the revised and updated GNVQ Engineering programme. Throughout we have adopted a common format and approach with numerous examples, problems and activities.

About GNVQ

General National Vocational Qualifications (GNVQ) are available in schools and colleges throughout England, Wales and Northern Ireland. The main aim of these 'vocational A-levels' is that of raising the status of vocational education in the UK within a new system of high quality vocational qualifications which can be taken as an alternative to the well-established General Certificate of Secondary Education (GCSE) and General Certificate of Educational Advanced Level (GCE A-level) qualifications.

With the advent of the latest Government initiative, *Curriculum 2000*, GNVQ awards have taken on an important new role in a highly flexible curriculum for students aged 16 to 19. It is now possible to mix vocational and non-vocational studies and many students will, for the first time, be able to try their hand at engineering whilst studying for more conventional A-level qualifications.

As well as acquiring the basic skills and body of knowledge that underpin a vocational area, all GNVQ students have to achieve a number of *key skills*. The attainment of both vocational *and* key skills provides a foundation from which students can progress either to further and higher education, or into employment with further training appropriate to the job concerned.

GNVQs, like their more vocationally focused NVQ counterparts, are unit-based qualifications. Each is made up of a number of units which can be assessed separately, and this allows credit accumulation throughout a course. A certificate can be obtained for each unit if necessary and credit transfer on some units can be made between qualifications. To make this possible, assessment is based on the unit rather than the qualification.

How to use this book

This book covers the four mandatory units that make up the GNVQ Intermediate Engineering programme. One chapter is devoted to each unit. Each chapter contains text, worked examples, 'test your knowledge' questions, activities, and review questions.

The worked examples will not only show you how to solve simple problems but they will also help put the subject matter into context with typical illustrative examples.

The 'test your knowledge' questions are interspersed with the text throughout the book. These questions allow you to check your understanding of the preceding text. They also provide you with an opportunity to reflect on what you have learned and consolidate this in manageable chunks.

Most 'test you knowledge' questions can be answered in only a few minutes and the necessary information, formulae, etc. can be gleaned from the surrounding text. Activities, on the other hand, require a significantly greater amount of time to complete. Furthermore, they often require additional library or resource area research coupled with access to computing and other information technology resources.

Activities make excellent vehicles for gathering the necessary evidence to demonstrate that you are competent in core skills.

Finally, here are a few general points worth noting:

- Allow regular time for reading – get into the habit of setting aside an hour, or two, at the weekend to take a second look at the topics that you have covered during the week.

- Make notes and file these away neatly for future reference – lists of facts, definitions and formulae are particularly useful for revision!

- Look out for the inter-relationship between subjects and units – you will find many ideas and a number of themes that crop up in different places and in different units. These can often help to reinforce your understanding.

- Don't expect to find all subjects and topics within the course equally interesting. There may be parts that, for a whole variety of reasons, don't immediately fire your enthusiasm. There is nothing unusual in this, however do remember that something that may not appear particularly useful now may become crucial at some point in the future!

- However difficult things seem to get – don't be tempted to give up! Engineering is not, in itself, a difficult subject, rather it is a subject that *demands* logical thinking and an approach in which each new concept builds upon those that have gone before.

- Don't be afraid to put your new ideas into practice. Engineering is about *doing* – get out there and *do* it!

Good luck with your GNVQ Engineering studies!

Mike Tooley

Unit 1 | Design and graphical communication

Summary

This unit is about the process of product design. It deals with all aspects of the process of designing an engineered product. The unit will also help you to develop your skills in design and graphical communication. Since designers make extensive use of drawings to present information, we shall begin the unit by introducing the different types of drawing that you are likely to meet in industry. You need to be able to read and understand these drawings and also develop skills in using both manual and CAD drawing techniques (see Unit 2). Within the unit we have used the relevant British Standards Institute (BSI) symbols and conventions. The full standards are very lengthy and expensive publications. However, low-cost, abridged editions are available for student use. These are: PP7307 Graphical Symbols for use in schools and colleges, and PP7308 Engineering Drawing Practice for Schools and Colleges. These are more than adequate for your requirements and you should either purchase copies or obtain a copy from your school or college library. This unit has links with many other units and in particular with Unit 3 (Make engineered products).

Communicating engineering information

Many ways of conveying information are used in the engineering industry. We can group these broadly as:

- the spoken word (oral communication)
- the written word
- the use of graphical representations (drawings, sketches, graphs, etc.).

The spoken word

The spoken word is widely used in the following situations:

- informal discussions either on the telephone or face to face
- formal presentations to groups of persons who all require the same information.

Where a group of persons all require the same information, a formal presentation must be used. On no account should information be 'passed down the line' from person to person. Errors are bound to creep in.

It is important to remember that the spoken word is easily forgotten and oral communication should be reinforced by:

- notes taken at the time
- tape recording the conversation
- a written summary. For example the published 'proceedings' of formal lectures and presentations. Another example is the 'press release' to ensure factual accuracy of information intended for the public.

Oral communication must be presented in a manner appropriate to the audience. It must be brief and to the point. The key facts must be emphasized so that they can be easily remembered. The presentation must be interesting so that the attention of the audience does not wander.

When communicating by the spoken word, it is as equally important to be a good listener as it is to be a good speaker. This applies to conversations between two or three people as well as to formal presentations.

Written communication

This is a more reliable method of communication since it provides a permanent record of the key information. The same information is available for all the persons who require it.

Anyone who has ever marked an English comprehension test will know that the same written passage can mean very different things to different persons. Therefore, care must be taken in preparing written information. To avoid confusion, the normal conventions of grammar and punctuation must be used. Words must be spelt correctly. Use a dictionary if you are uncertain. If you are using a word processing package use the spell-checker. However, take care, many software packages originate in the USA and the spell-checker may reflect this.

Never use jargon terms and acronyms unless you are sure that the persons reading the message are as equally familiar with them as is the writer.

An engineer often has to write notes, memoranda, and reports. He or she often has to maintain logbooks and complete service sheets. An engineer may also have to communicate with other engineers, suppliers and customers by letter. Practice in the writing of clear and concise messages is of great importance.

Graphical communication

Engineers rely heavily upon graphical methods of communication. Drawings and charts produced to international standards using international symbols and conventions suffer no language barriers. They are not liable to be misinterpreted by translation errors. Graphical communication does not replace spoken and written communication. It is used to simplify, reinforce and complement other means of communication.

Design office

A designer will 'conceptualize' the design for a new product using free hand sketches and often simple cardboard models. These preliminary stages will include engineering the design of the product so that it will function correctly, and styling the product so that it is attractive to the customer.

This can be made easier by the use of advanced computer aided design (CAD) software. This advanced software enables three-dimensional representations to be prepared by 'wire-frame' or solid modelling. Such images can be rotated on the screen so that viewing is possible from all angles. Modifications to the design can be made at the touch of a key. Further, the function of the design can be checked by simulation techniques, before embarking on expensive prototype manufacture.

Once these preliminaries have been achieved, the design team will need to consult with representatives of the customer, with maintenance and field service engineers, and with the various bodies responsible for type approval. These may include insurance engineers and Local Authority environmental engineers.

Finally the design will need to be proved by the manufacture and testing of a full-scale prototype.

Drawing office

Once the design has been approved, technical drawings and specifications will be produced by draughtspersons. These drawings and specifications will vary depending upon for whom they are intended. The manufacturing engineer will want flow charts, and orthographic detail and general arrangement drawings. The service engineer will want circuit diagrams and exploded views. Industrial customers will want installation drawings and operating data. Drawings and specifications are also required for other purposes.

Marketing

Formal engineering drawings are not appropriate for most customers. Usually, sales material is produced by graphic artists who combine colour photographs with performance data, often in the form of graphs; for example the glossy brochures available in car showrooms. This type of material is closely geared to the potential customer. The pictorial material is usually enhanced by a graphic artist using airbrush techniques.

Estimating, costing and planning

These functions are closely linked and involve both engineers and accountants. Fully detailed drawings and specifications are required, together with block diagrams, flow diagrams and spreadsheets. The cost of every component and operation is estimated so that a provisional selling price can be arrived at. Cost targets (budgets) are set and actual production costs are constantly compared with these targets. If the targets are not achieved or improved upon, then either the selling price has to be increased or the manufacturing process modified. Planning involves the economical loading of the work onto the production shops so that the machines are kept fully employed and the work flows smoothly so that delivery dates are achieved and customer goodwill is maintained.

Manufacturing

The workshops will require general arrangement drawings and detail drawings. The former show the assembled unit and list all the parts that are to be manufactured together with the identification numbers of the detail drawings required. Also shown and listed will be all the standard 'bought in' components such as nuts and bolts. The detail drawings show all the information needed to manufacture the individual components; that is, the material to be used, the shape of the components, the dimensions of the component and any heat treatment or special surface finishing treatments. Figure 1.1 shows an example of a General Arrangement (GA) drawing and Figure 1.2 shows an example of a detail drawing.

In addition, assembly drawings will be required. These are often confused with general arrangement drawings that can fulfil this function for very simple assemblies. For more complex assemblies, 'exploded' drawings are often used together with flow charts showing the sequence in which the components should be fitted together.

Service/maintenance

Pictorial drawings (isometric or oblique) are often used to show the various lubrication points on a machine, the type of lubricant to be used and the frequency of lubrication. These pictorial drawings are often 'exploded' and list the spare parts numbers, as shown in Figure 1.3. Such drawings also show the relationships between the parts to aid dismantling and assembly.

Activity 1.1

Compare and contrast the advantages and limitations of:

(a)   oral communication
(b)   written communication
(c)   graphical communication.

Present your work in the form of a word-processed script and three-minute talk recorded on cassette tape to be broadcast by a local radio station.

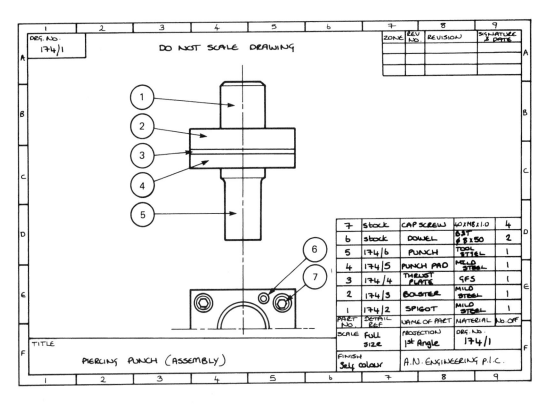

Figure 1.1 *General Arrangement (GA) drawing*

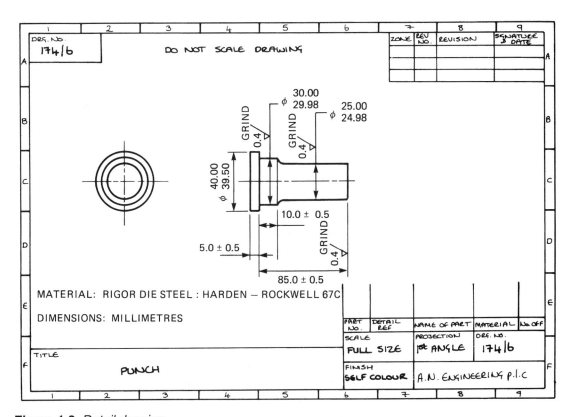

Figure 1.2 *Detail drawing*

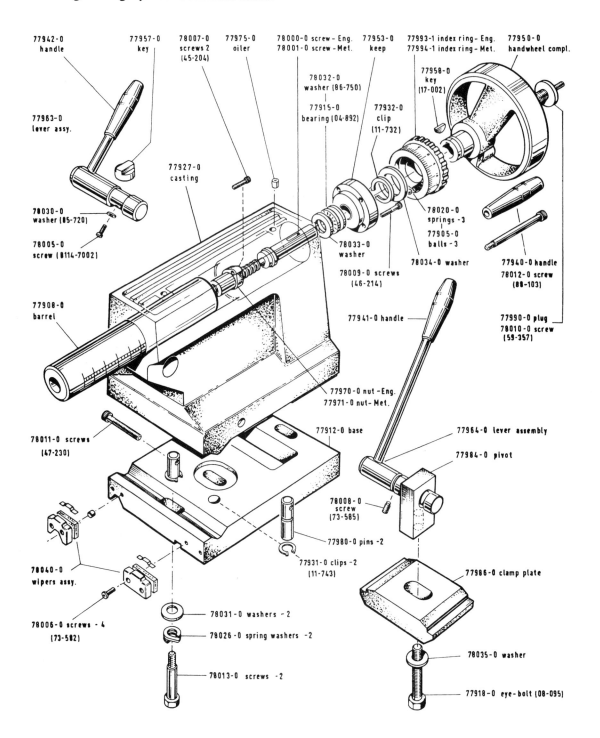

Figure 1.3 *Exploded view and parts list for the tailstock assembly*

Graphical methods for engineering information

Having now established the need for communicating engineering information, let's look at the various methods of graphical communication available. We can broadly divide engineering information into two categories: that which is mathematically based, and that which is technically based. Let's be logical and look at the former first.

Mathematical data

I expect that you will already have met most of the methods of expressing mathematical data by means of graphs. Nevertheless, let's revise the techniques available.

Line graphs

Just as engineering drawings are used as a clear and convenient way of describing complex components and assemblies, so can graphs be used to give a clear and convenient picture of the mathematical relationships between engineering and scientific quantities. Figure 1.4(a) shows a graph of the relationship between distance S and time t for the mathematical expression $S = \frac{1}{2} a\, t^2$ where the acceleration $a = 10$ m/s^2.

In this instance it is correct to use a continuous curve to connect the points plotted. Not only do these points lie on the curve, but every corresponding value of S and t between the points plotted also lie on the curve.

However, this is not true for every type of line graph. Figure 1.4(b) shows a graph relating speed and distance for a journey. From A to B the vehicle is accelerating. From B to C the vehicle is travelling at a constant speed. From C to D the vehicle is decelerating (slowing down). In this example it is correct to join the points by straight lines This is because each stage of the journey is represented by a linear mathematical expression which is unrelated to the previous stage of the journey and unrelated to the following stage of the journey.

Histograms

Consider the student intake of a college over a number of years. The students enrol in September and leave in the following July. Between these times there is a negligible change in the number of students attending. Therefore to plot the enrolments for each September and connect these by a flowing curve between the points would be incorrect. Such graphs would imply that there is a continuous change in student numbers between one September and the next and that the change satisfies a mathematical equation.

It would also be incorrect to connect the plotted points by straight lines. This would imply that, although the points plotted are not following a mathematical expression, there is some continuous change in the number of students attending between one September and the next.

The correct way to show this sort of data is by means of a histogram

1. Given the formula
    $v = 100 + (a\,t)$ where:
    v = velocity in m/s
    a = acceleration in m/s^2
    t = time in s
    plot the graph relating velocity
    to time when $a = 12.5$ m/s^2,
    15 m/s^2, and 22.5 m/s^2 for
    values of t between 0 and 4 s.
    Label your axes clearly.

2. From the graph determine the
    difference between the
    velocities when $a = 12.5$ m/s^2
    and $a = 22.5$ m/s^2 at $t = 2$ s.

3. Draw a graph showing the
    following furnace temperature
    temperatures taken at hourly
    intervals: 620°C, 710°C,
    750°C, 730°C, 790°C,
    720°C, 650°C, 525°C.
    For what approximate period
    of time does the furnace
    temperature exceed 700°C?

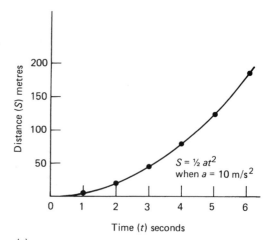

(a)

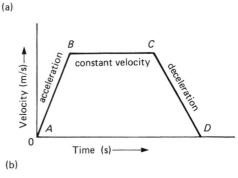

(b)

Figure 1.4 *Line graphs*

as shown in Figure 1.5. This clearly compares the attendances for each college year. At the same time it shows that each year's attendance stands alone and is unrelated to the previous year and unrelated to the following year.

The total number of machines
manufactured by a small
company each year is as follows:

1990	250
1991	350
1992	325
1993	375
1994	290
1995	180
1996	215

Draw a histogram to compare
these output quantities.

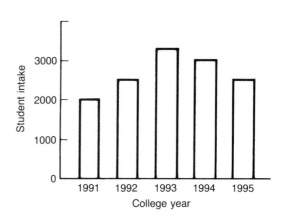

Figure 1.5 *A histogram*

Bar charts

These are also used for displaying statistical data, but are usually plotted horizontally. They are widely used for recording work in progress, as shown in Figure 1.6.

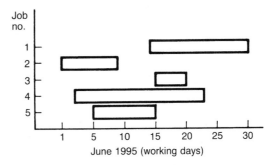

Figure 1.6 *A bar chart*

Ideographs (pictograms)

These are frequently used to present statistical data to the general public. A typical example of the number of cars using a car park is shown in Figure 1.7. In this example each symbol represents 1000 cars. Therefore in 1994, 3000 cars used the car park each week (1000 cars multiplied by 3 symbols).

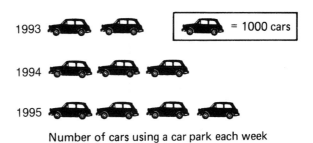

Number of cars using a car park each week

Figure 1.7 *An ideograph*

Test your knowledge 1.3

Use an ideograph to show the following production statistics for a CD manufacturer. Each symbol should represent 10,000 units.

January	20,000
February	22,000
March	30,000
April	45,000
May	60,000
June	55,000

Pie-charts

These are used to show how a total quantity is divided up into its individual parts. Let's look at Figure 1.8(a). Since a complete circle is 360°, we can represent 25% of a complete circle as $360° \times 25/100 = 90°$. Figure 1.8(b) shows how the expenditure of a company can be represented by a pie-chart.

Draw a pie chart for the following data representing the per unit breakdown of costs for a product:

Labour costs   40%
Material costs   25%
Overhead costs   12.5%
Profit   22.5%

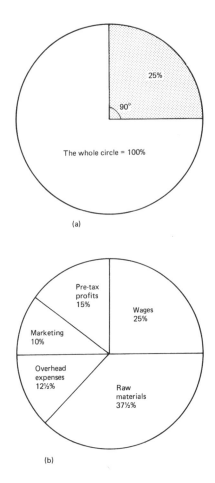

(a)

(b)

Figure 1.8 *Pie charts*

Activity 1.2

(a)   Figure 1.9 shows a graph relating electrical potential and current for a particular circuit.

   (i) State the magnitude of the current in amperes when the potential is 30 V.
   (ii) State the magnitude of the potential in volts when the current is 4 A.

(b)   (i) Name the type of graph shown in Figure 1.10.
   (ii) State how many students attended in 1990, 1991, and 1994.

(c)   Draw a histogram to show the number of attendances shown in Figure 1.10.

(d)   The breakdown of costs for a product is: Materials = £500, Labour = £1,000, Overheads = £250, Profit = £250. Represent this information in the form of a pie chart.

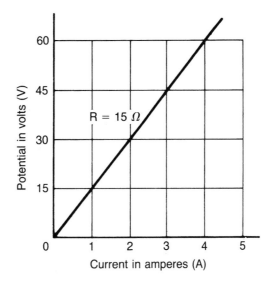

Figure 1.9 *See Activity 1.2*

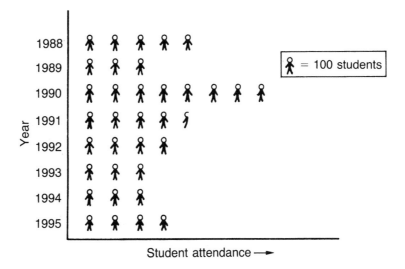

Figure 1.10 *See Activity 1.2*

Technical information and drawings

Like the graphs that we have just considered, there are many different ways of representing and communicating technical information. To avoid confusion such information should make use of nationally and internationally recognized symbols, conventions and abbreviations. These are listed and their use explained in the appropriate British Standards. Such standards are lengthy and costly. Low-cost summaries of these standards are available for students. These abridged editions are: PP7307 Graphical Symbols for Use in Schools and Colleges; and PP7308 Engineering Drawing Practice for Schools and Colleges (abridged from BS 308).

Use a block diagram to show how the engine drives the road wheels of a car via such elements as the differential, gearbox, and clutch.

Block diagrams

These show the relationship between the various elements of a system. Figure 1.11 shows the block diagram for a simple radio receiver. These sorts of diagrams are used in the initial stages of conceptualizing a design.

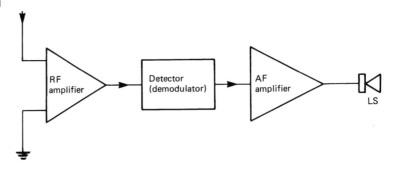

Figure 1.11 *A block diagram*

Flow diagrams

These are used to test the logic of a sequence of events. They are used for a variety of purposes by computer software engineers when planning a new program, and by production engineers in working out the best sequence of operations in which to manufacture a component. Figure 1.12 shows a flow chart for drilling a hole. The shape of the 'boxes' used in this flow chart have particular meanings, as shown in Figure 1.13. For the complete set of 'boxes', their meanings and uses see British Standard PP3707.

Your bicycle tyre is flat. Draw a flow diagram for checking the inner-tube, locating and repairing a puncture (if there is one), or replacing a faulty valve. Figure 1.14 shows you how to start the flow chart.

Circuit diagrams

These are used to show the functional relationships between the components in a circuit. The components are represented by symbols and their position in the circuit diagram does not represent their actual position in the final assembly. Circuit diagrams are also referred to as schematic diagrams or even schematic circuit diagrams.

Figure 1.15(a) shows the circuit for an electronic filter unit using standard component symbols. Figure 1.15(b) shows a layout diagram with the components correctly positioned. Figure 1.15(c) shows the printed circuit board. This is also called a track diagram or a wiring diagram.

However, it is more usual to use the term wiring diagram where the components are hard wired, as in the wiring up of a building or the manufacture of a control cubicle. Architects use circuit diagrams to show the electrical installation of buildings. They also provide installation drawings to show where the components are to be sited. They may also provide a wiring diagram to show how the cables are to be routed to and between the components. The symbols used in

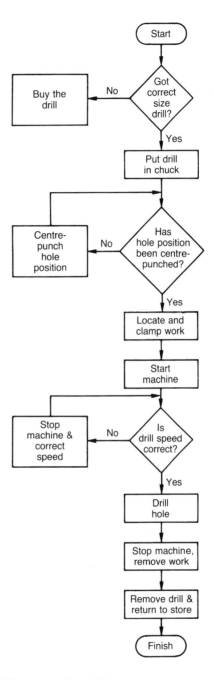

Figure 1.12 *Flow chart for drilling a hole*

architectural installation drawings and wiring diagrams are not the same as those used in circuit diagrams. Examples of architectural and topographical electrical component symbols are shown in BS: PP7307.

Schematic circuit diagrams are also used to represent pneumatic (compressed air) circuits and hydraulic circuits. Pneumatic circuits and hydraulic circuits share the same symbols. You can tell which circuit is which because pneumatic circuits should have open arrow heads, whilst hydraulic circuits should have solid arrowheads. Also,

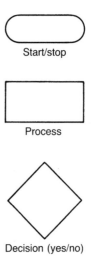

Figure 1.13 *Some common flow chart symbols*

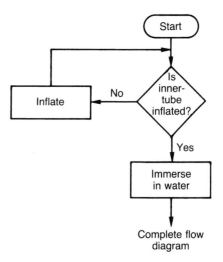

Figure 1.14 *See Test Your Knowledge 1.6*

pneumatic circuits exhaust to the atmosphere, whilst hydraulic circuits have to have a return path to the oil reservoir. Figure 1.16 shows a typical hydraulic circuit.

Just as electrical circuit diagrams may have corresponding installation and wiring diagrams, so do hydraulic, pneumatic and plumbing circuits. Only this time the wiring diagram becomes a pipe-work diagram. A plumbing example is shown in Figure 1.17. As you may not be familiar with the symbols, I have named them for you. Normally this is not necessary and the symbols are recognized by their shapes.

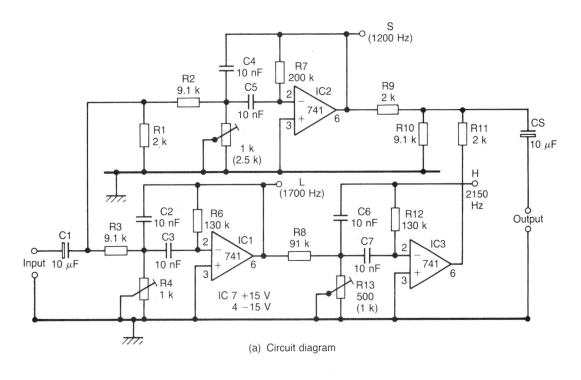

(a) Circuit diagram

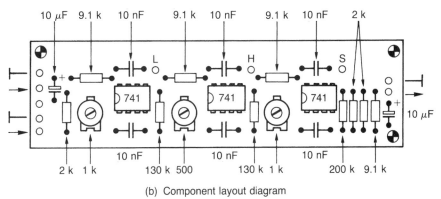

(b) Component layout diagram

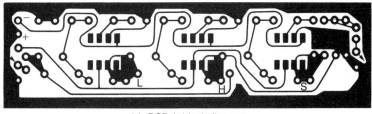

(c) PCB (wiring) diagram

Figure 1.15 *A typical electronic circuit diagram with corresponding layout diagram and PCB (wiring) diagram*

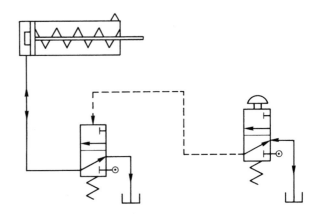

Figure 1.16 *A typical hydraulic circuit*

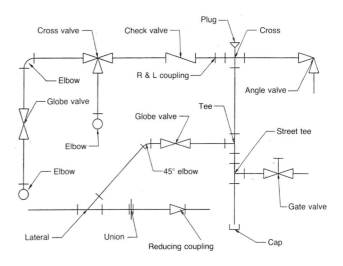

(a) Circuit diagram (schematic)

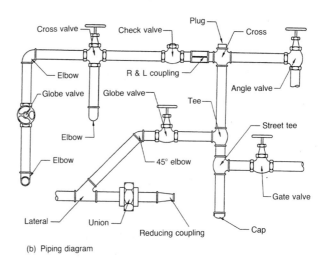

(b) Piping diagram

Figure 1.17 *A typical plumbing circuit with corresponding piping diagram*

General Arrangement (GA) drawings

Figure 1.18 shows the layout of a typical drawing sheet. To save time these are printed to a standardized layout for a particular company, ready for the draughtsperson to add the drawing and complete the boxes and tables.

The basic information found on most drawing sheets consists of:

- the drawing number and name of the company
- the title and issue details
- scale
- method of projection (first or third angle)
- initials of persons responsible for: drawing, checking, approving, tracing, etc. together with the appropriate dates
- unit(s) of measurement (inches or millimetres) and general tolerances
- material and finish
- copyright and standards reference
- guidance notes such as: 'do not scale'
- reference grids so that 'zones' on the drawing sheet can be quickly found
- modifications table for alterations which are reference related to the issue number on the drawing and identified by the means of the reference grid.

The following additional information may also be included:

- fold marks
- centre marks for camera alignment when microfilming
- line scale, so that the true size is not lost when enlarging or reducing copies
- trim marks
- orientation marks.

Figure 1.19 shows a typical General Arrangement (GA) drawing. This shows as many of the features listed above as are appropriate for this drawing. It shows all the components correctly assembled together. Dimensions are not usually given on GA drawings although, sometimes, overall dimensions will be given for reference when the GA drawing is of a large assembly drawn to a reduced scale.

The GA drawing shows all the parts. These are listed in a table together with the quantities required. Manufacturers' catalogue references are also given for bought-in components. The detail drawing numbers are also included for components that have to be manufactured as special items.

Detail drawings

As the name implies, detail drawings provide all the details required to make the component shown on the drawing. Referring to Figure 1.19, we see from the table that the detail drawing for the punch has the reference number 174/6. Figure 1.20 shows this detail drawing. In this instance, the drawing provides the following information:

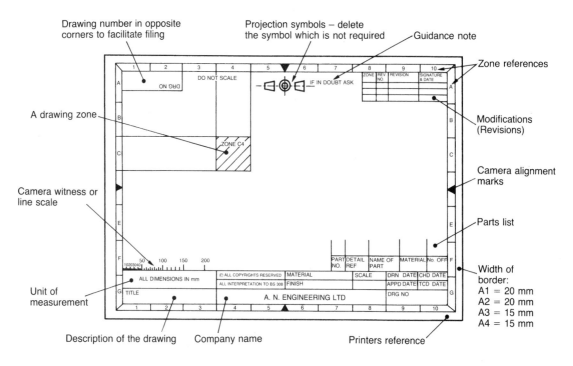

Figure 1.18 *Layout of a typical drawing sheet*

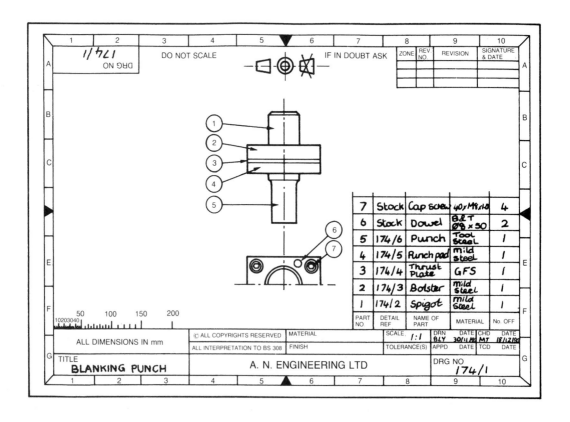

Figure 1.19 *A typical General Arrangement (GA) drawing*

- the shape of the punch
- the dimensions of the punch and the manufacturing tolerances
- the material from which the punch is to be made and its subsequent heat treatment
- the unit of measurement (millimetre)
- the projection (first angle)
- the finish
- the guidance note 'do not scale drawing'
- the name of the company
- the name of the draughtsperson
- the name of the person checking the drawing.

The amount of information given will depend upon the job. Drawings for a critical aircraft component will be much more fully detailed than a drawing for a wheelbarrow component.

Data storage

There are many different ways of storing technical data.

- *Tracing linen* This was the traditional material for making technical drawings. It was strong and durable and stood up well to the effects of the ultra-violet arc lamps used for making 'blue-print' copies. They got their name from the fact that the print appeared as white lines on a blue background.

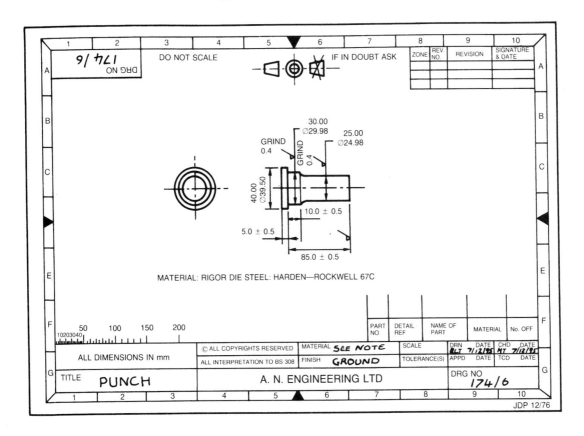

Figure 1.20 *A typical detail drawing*

- *Tracing paper* This is widely used in conjunction with manual drawing techniques. It is cheap and readily available. It is also easy to draw on. Unfortunately the paper becomes brittle with age and requires careful handling. Therefore it is not suitable where print copies have to be made frequently.
- *Tracing film* This is tough plastic film that is shiny on one side and matt on the other. You draw on the matt surface. No special techniques are required in its use and it stands up to repeated handling without deterioration. It is more expensive than tracing paper.
- *Microfilm* The storage of full size 'negatives', as the tracings are called, when produced on linen, paper or film takes up a lot of room. These large drawings can be reduced photographically onto 16 mm or 35 mm film stock for storage. This saves considerable space. The microfilm copies can be projection printed (enlarged) full-size when required for issue.
- *Microfiche* Libraries, offices and stores use microfiche systems. Data is stored photographically in a grid of frames on a large rectangle of film. A desktop viewer is used to select and enlarge a single frame. The frame is then projected onto a rear projection screen for easy reading. This system is more likely to be used for storing literal and numerical data than for drawings.
- *Electronic* Computer aided drawing and design (CAD) has rapidly taken over from manual drawing and tracing. The advantages of CAD will be discussed later in this Unit. CAD software is used in conjunction with a computer and the 'drawing' produced on the computer screen is saved in a computer file on disk. This is the most economical method of data storage. Many very complex drawings can be saved on a single floppy disk. When required the file can be recalled for immediate viewing on a computer screen and hard copy can be printed out to any desired scale at the touch of a key.

Activity 1.3

(a)   Compare the selection of sample printed drawing sheets provided by your tutor with the model sheet shown in Figure 1.18. Comment on any differences.

(b)   Examine the sample General Arrangement (GA) drawings provided by your tutor and compare them with the GA drawing shown in Figure 1.19. Comment on any differences.

(c)   Examine the sample detail drawings provided by your tutor and compare them with the model detail drawing shown in Figure 1.20. Again, comment on any differences.

Activity 1.4

Draw a flow diagram for connecting a conventional 13 A plug to the three-core AC mains lead fitted to an electrical appliance. Don't forget to allow for decision making, e.g. is the fuse rating correct? Produce your flowchart in the form of an overhead projector transparency.

Activity 1.5

Draw a block diagram showing the arrangement of the braking system of a conventional passenger car. Prepare your block diagram using a computer drawing or CAD package.

Activity 1.6

List the basic equipment required for making a technical drawing manually and explain what each item is used for, Present your answer in the form of a word processed instruction card suitable for someone who is a complete newcomer to technical drawing.

Engineering drawing techniques

Engineering drawings can be produced either manually or on a computer using suitable CAD software. Drawings produced manually can range from freehand sketches to formally prepared drawings produced with the aid of a drawing board and conventional drawing instruments.

The choice of technique is dependent upon a number of factors such as:

- *Speed* How much time can be allowed for producing the drawing. How soon the drawing can be commenced.
- *Media* The choice will depend upon the equipment available (e.g. CAD or conventional drawing board and instruments) and the skill of the person producing the drawing.
- *Complexity* The amount of detail required and the anticipated amount and frequency of development modifications.
- *Cost* Engineering drawings are not works of art and have no intrinsic value. They are only a means to an end and should be produced as cheaply as possible. Both initial and on-going costs must be considered.

- *Presentation*   This will depend upon who will see/use the drawings. Non-technical people can visualize pictorial representations better than orthographic drawings.

Nowadays technical drawings are increasingly produced using computer aided drawing techniques (CAD). Developments in software and personal computers have reduced the cost of CAD and made it more powerful. At the same time, it has become more 'user friendly'. Computer aided drawing does not require the high physical skill required for manual drawing which takes years of practice to achieve. It also has a number of other advantages over manual drawing. Let's consider some of these advantages:

- *Accuracy*   Dimensional control does not depend upon the draughtsperson's eyesight.
- *Radii*   These can be made to blend with straight lines automatically.
- *Repetitive features*   For example, holes round a pitch circle do not have to be individually drawn but can be easily produced automatically by 'mirror imaging'. Again, some repeated, complex feature need only be drawn once and saved as a matrix. It can then be called up from the computer memory at each point in the drawing where it appears at the touch of a key.
- *Editing*   Every time you erase and alter a manually produced drawing on tracing paper or plastic film the surface of the drawing is increasingly damaged. On a computer you can delete and redraw as often as you like with no ill effects.
- *Storage*   No matter how large and complex the drawing, it can be stored digitally on floppy disk. Copies can be taken and transmitted between factories without errors or deterioration.
- *Prints*   Hard copy can be produced accurately and easily on flat bed or drum plotters and to any scale. Colour prints can also be made.

Engineering drawings such as General Arrangement drawings and detail drawings are produced by a technique called orthographic drawing using the conventions set out in BS 308. Since I will be asking you to make orthographic drawings from more easily recognized pictorial drawings, I will start by introducing you to the two pictorial techniques widely used by draughtspersons.

Oblique drawing

Figure 1.21 shows a simple oblique drawing. The front view (elevation) is drawn to true shape and size. Therefore this view should be chosen so as to include any circles or arcs so that these can be drawn with compasses. The lines forming the side views appear to travel away from you, so these are called 'receders'. They are drawn at 45° to the horizontal using a 45° set-square. They may be drawn full length as in cavalier oblique drawing or they may be drawn half-length as in cabinet oblique drawing. This latter method gives a more realistic representation, and is the one you and I will be using.

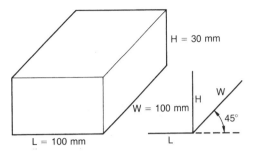

Cavalier oblique projection

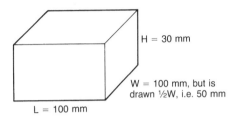

Cabinet oblique projection

Figure 1.21 *A simple oblique drawing*

Isometric drawing

Figure 1.22(a) shows an isometric drawing of our previous box. To be strictly accurate, the vertical lines should be drawn true length and the receders should be drawn to a special isometric scale. However this sort of accuracy is rarely required and, for all practical purposes, we draw all the lines full size. As you can see, the receders are drawn at 30° to the horizontal for both the elevation and the end view.

Although an isometric drawing is more pleasing to the eye, it has the disadvantage that all circles and arcs have to be constructed. They cannot be drawn with compasses. Figures 1.22(b), 1.22(c) and 1.22 (d) show you how to construct an isometric curve. You could have used this technique in Test your knowledge 1.7 to draw the circle on the side of the box drawn in oblique projection.

First we draw the required circle. Then we draw a grid over it as shown in Figure 1.22(b). Next number or letter the points where the circle cuts the grid as shown. Now draw the grid on the side elevation of the box and step off the points where the circle cuts the grid with your compasses as shown in Figure 1.22(c). All that remains is to join up the dots and you have an isometric circle as shown in Figure 1.22(d).

Another way of drawing isometric circles and curves is the 'four-arcs' method. This does not produce true curves but they are near enough for all practical purposes and quicker and easier than the previous method for constructing true curves. The steps are shown in Figure 1.23.

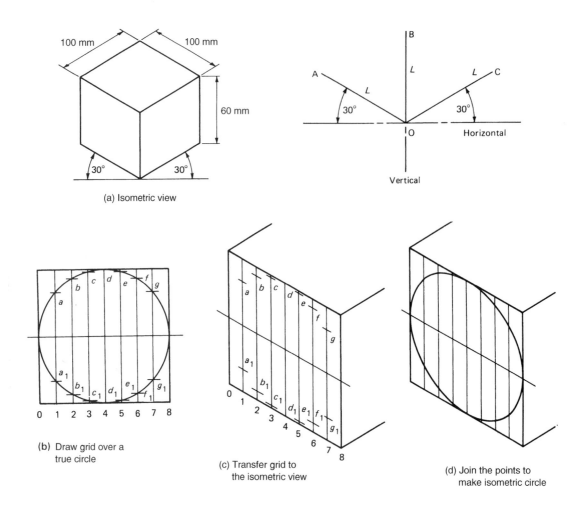

Figure 1.22 *Isometric drawing*

(a) Draw, full size, an isometric view of the box shown in Figure 1.22. Isometric ruled paper will be of great assistance if you can obtain some!

(b) Draw a 50 mm diameter isometric circle on the upper (top) face of the box. Remember that Figure 1.22 shows it on the side of the box.

- Join points B and E, as shown in Figure 1.23(b). The line BE cuts the line GC at the point J. The point J is the centre of the first arc. With radius BJ set your compass to strike the first arc as shown.
- Join the points A and F, as shown in Figure 1.23(c). The line AF cuts the line GC at the point K. The point K is the centre of the second arc. With radius KF set your compasses to strike the second arc as shown. If your drawing is accurate both arcs should have the same radius.
- With centre A and radius AF or AD strike the third arc, as shown in Figure 1.23(d).
- With centre E and radius EH or EB strike the fourth and final arc, as shown in Figure 1.23(e). If your drawing is accurate, arcs 3 and 4 should have the same radius.

The Test your knowledge exercises will provide you with some practise with isometric drawings. You should complete these before moving on to the next section.

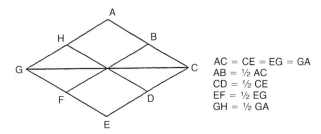

AC = CE = EG = GA
AB = ½ AC
CD = ½ CE
EF = ½ EG
GH = ½ GA

(a) Draw an isometric grid of appropriate size

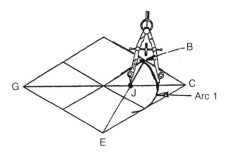

(b) Construct the 1st arc using a compass located as shown

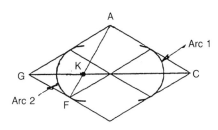

(c) Draw the 2nd arc using the construction process shown

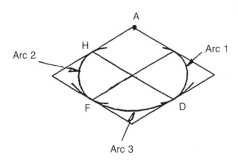

(d) Draw the 3rd arc through the points shown

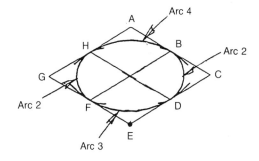

(e) Complete the process drawing the 4th arc from the opposite corner

Figure 1.23 *The 'four-arcs' method*

Orthographic drawing

General Arrangement (GA) and detail drawings are produced by the use of a drawing technique called orthographic projection. This is used to represent three-dimensional solids on the two-dimensional surface of a sheet of drawing paper so that all the dimensions are true length and all the surfaces are true shape. To achieve this when surfaces are inclined to the vertical or the horizontal we have to use auxiliary views, but more about these later. Let's keep things simple for the moment.

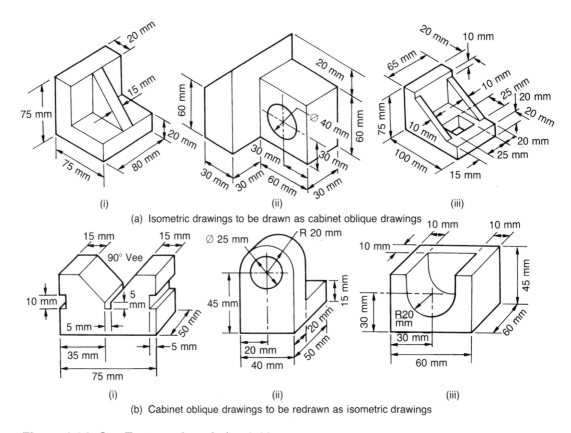

(a) Isometric drawings to be drawn as cabinet oblique drawings

(b) Cabinet oblique drawings to be redrawn as isometric drawings

Figure 1.24 *See Test your knowledge 1.10*

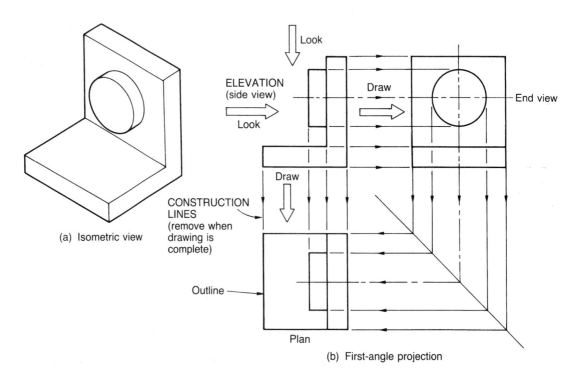

(a) Isometric view

(b) First-angle projection

Figure 1.25 *An isometric view and its corresponding first-angle projection*

Test your knowledge 1.10

(a) Figure 1.24(a) shows some further examples of isometric drawings. Redraw them as cabinet oblique drawings.

(b) Figure 1.24(b) shows some further examples of cabinet oblique drawings. Redraw them as isometric drawings. Any circles and arcs on the vertical surfaces should be drawn using the grid construction method. Any arcs and circles on the horizontal (plan) surfaces should be drawn using the 'four-arcs method'.

Test your knowledge 1.11

Figure 1.24 shows some components using pictorial projections. Redraw each of these drawings in first-angle orthographic projection. To start you off the first one has been drawn for you (see Figure 1.27). Note how the end view has been positioned so that you can see the web.

First-angle projection

Figure 1.25(a) shows a simple component drawn in isometric projection. Figure 1.25(b) shows the same component as an orthographic drawing. This time we make no attempt to represent the component pictorially. Each view of each face is drawn out separately either full size or to the same scale. What is important is how we position the various views as this determines how we 'read' the drawing.

Engineers use two orthographic drawing techniques, either first-angle or third-angle projection. The former is called 'English' projection and the latter is known as 'American' projection. The drawing in Figure 1.25 is in first-angle projection. The views are arranged as follows:

- *Elevation* This is the main view from which all the other views are positioned. You look directly at the side of the component and draw what you see.
- *Plan* To draw this, you look directly down on the top of the component and draw what you see below the elevation.
- *End view* This is sometimes called an 'end elevation'. To draw this you look directly at the end of the component and draw what you see at the opposite end of the elevation. There may be two end views, one at each end of the elevation, or there may be only one end view if this is all that is required to completely depict the component. Figure 1.25 requires only one end view. When there is only one end view this can be placed at either end of the elevation depending upon which position gives the greater clarity and ease of interpretation. Whichever end is chosen, the rules for drawing this view must be obeyed.

Use faint construction lines to produce the drawing, as shown in Figure 1.25(b). When these are complete, 'line-out' the outline more heavily. Carefully remove the construction lines to leave the drawing uncluttered, thus improving the clarity. Sometimes, in examinations, you are asked to leave the construction lines so that the examiner can see how you have arrived at your answer. Figure 1.26 shows the finished drawing.

Third-angle projection

Figure 1.28 shows the same component, but this time I have drawn it in third-angle projection for you.

- *Elevation* Again I have started with the elevation or side view of the component and, as you can see, there is no difference.
- *Plan* Again we look down on top of the component to see what the plan view looks like. However, this time, we draw the plan view above the elevation. That is, in third-angle projection we draw all the views from where we look.
- *End view* Note how the position of the end view is reversed compared with first-angle projection. This is because, like the plan view, we draw the end views at the same end from which we look at the component.

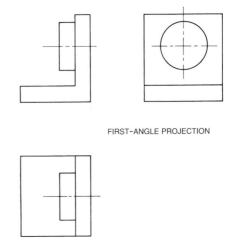

FIRST-ANGLE PROJECTION

Figure 1.26 *Completed first-angle projection*

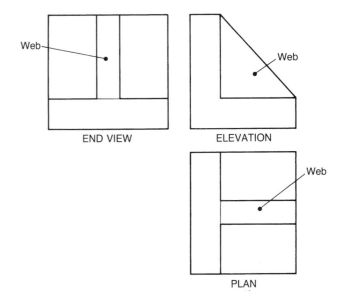

END VIEW

ELEVATION

PLAN

Web

Web

Web

Figure 1.27 *See Test your knowledge 1.11*

Again use faint construction lines to produce the drawing as shown in Figure 1.28. Then 'line-in' the outline more heavily and carefully remove the construction lines for clarity, unless you have been instructed otherwise. Figure 1.29 shows the finished drawing in third-angle projection.

Auxiliary views

In addition to the main views on which we have just been working, we sometimes have to use auxiliary views. We use auxiliary views when we cannot show the true outline of the component or a feature of the component in one of the main views; for example, when a surface of the component is inclined as shown in Figure 1.33.

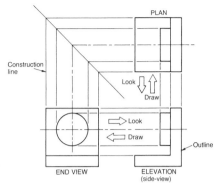

Figure 1.28 *Third-angle projection*

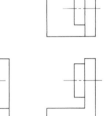

Figure 1.29 *Completed third-angle projection*

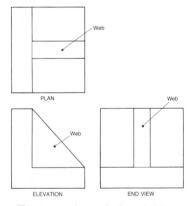

Figure 1.30 *See Test your knowledge 1.12*

Test your knowledge 1.13

(a) Figure 1.31 shows some components drawn in first-angle projection and some in third-angle projection. Note that not all the possible views have been shown (only as many are shown as are actually needed). State which is first-angle and which is third-angle.

(b) Two of the drawings in Figure 1.31 are standard symbols for indicating whether a drawing is in first-angle or whether it is in third-angle. Which drawings do you think show these two symbols?

Test your knowledge 1.14

Figure 1.32 shows pictorial views of some more solid objects.

(a) Draw these objects in first-angle orthographic projection and label the views.

(b) Draw these objects in third-angle orthographic projection and label the views.

Activity 1.7

Draw the component shown in Figure 1.34 in isometric projection. Each square has a side length of 10 mm. Also draw the component in:

(a) first-angle orthographic projection (only two views required)

(b) third-angle orthographic projection (only two views required)

(c) cabinet oblique projection.

Present your results in the form of a portfolio of drawings. Clearly mark each drawing with the projection used.

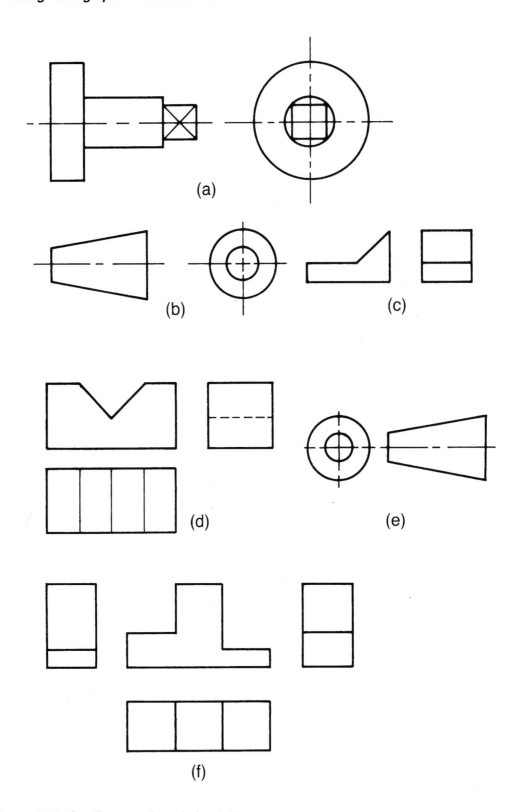

Figure 1.31 *See Test your knowledge 1.13*

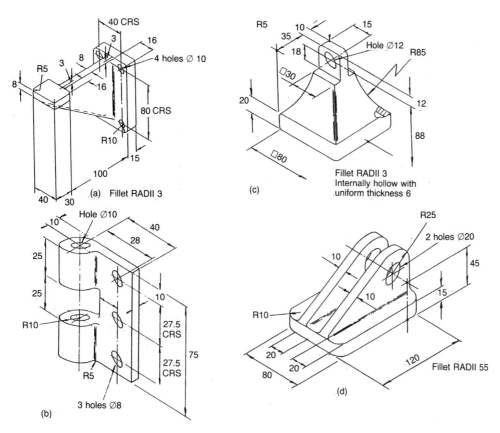

Figure 1.32 *See Test your knowledge 1.14*

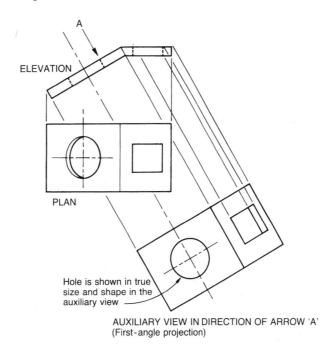

AUXILIARY VIEW IN DIRECTION OF ARROW 'A'
(First-angle projection)

Figure 1.33 *An auxiliary view*

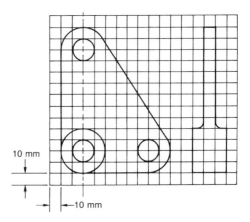

10 mm

10 mm

Figure 1.34 *See Activity 1.7*

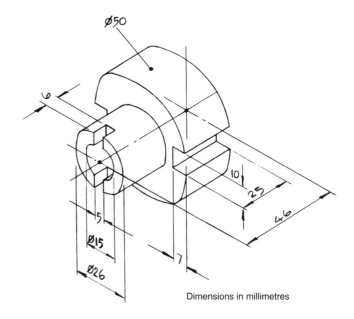

Dimensions in millimetres

Figure 1.35 *A dimensioned drawing*

Production of engineering drawings

Standard conventions are used in order to avoid having to draw out, in detail, common features in frequent use. Figure 1.35 shows a typical dimensioned engineering drawing. Figure 1.36(a) shows a pictorial representation of a screw thread. Figure 1.36(b) shows the convention for a screw thread. The convention for the screw thread is much the quicker and easier to draw.

All engineering drawings should be produced using appropriate drawing standards and conventions for the following reasons:

*   *Time* It speeds up the drawing process by making life easier for the draughtsperson as indicated above. This reduces costs and also reduces the 'lead-time' required to get a new product into production.

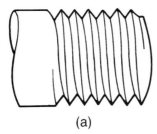 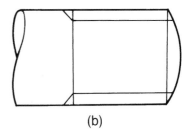

(a)   (b)

Figure 1.36 *Screw threads*

- *Appearance* The consistent use standards and conventions not only serves to make your drawings look more professional but it also improves the 'image' of yourself and your company. Badly presented drawings can send out the wrong messages and can call your competence into question.
- *Portability* Drawings produced to international standards and conventions can be read and correctly interpreted by any trained engineer, anywhere in this country or abroad. This avoids misunderstandings that could lead to expensive and complex components and assemblies being scrapped and dangerous situations arising. (Note that problems can still sometimes occur due to text and written notes that may be language dependent.)

Drawing conventions used by engineers in the UK are specified in BS 308. This is produced in three parts:

- Part 1 General principles.
- Part 2 Dimensioning and tolerancing of size.
- Part 3 Geometrical tolerancing.

These are all 'harmonized' with their appropriate ISO counterparts (ISO = International Standards Organization).

As has been stated earlier, you should get yourself a copy of the British Standards Institution's publication PP7308: Engineering Drawing Practice for Schools and Colleges. Also useful is PP7307: Graphical Symbols for Use in Schools and Colleges.

PP7307 contains symbols for use in electrical, electronic, pneumatic and hydraulic schematic circuit diagrams and many other useful symbols.

Since they are intended for students, both PP7307 and PP7308 are sold at very low prices compared with the full standards. If you cannot obtain a personal copy it is worth locating one in your school or college library.

Other British Standards of importance to engineering draughtspersons are:

- BS 4500: ISO Limits and Fits (these are used by mechanical and electronics engineers).
- BS 3939: Graphical symbols for electrical power, telecommunications and electronics diagrams.
- BS 2197: Specifications for graphical symbols used in diagrams for fluid power systems and components.

Planning the drawing

Before we start the actual drawing and lay pencil to paper we should plan what we are going to do. This saves having to alter the drawing or even starting again later on. We have to decide whether the drawing is to be pictorial, orthographic or schematic. If orthographic we have to decide on the projection we are going to use. We also have to decide whether we need a formal drawing or whether a freehand sketch is all that is required. If a formal drawing is needed then we have to decide whether to use manual techniques or CAD.

Paper size

When you start to plan your drawing you have to decide on the paper size that you are going to use. Engineering drawings are usually produced on 'A' size paper. Paper size A0 is approximately one square metre in area and is the basis of the system. Size A1 is half the area of size A0, size A2 is half the area of size A1 and so on down to size A4. Smaller sizes are available but they are not used for drawing. All the 'A' size sheets have their sides in the ratio of 1:1.2. This gives the following paper sizes (see Figure 1.37):

A0 = 841 mm × 1189 mm
A1 = 594 mm × 841 mm
A2 = 420 mm × 594 mm
A3 = 297 mm × 420 mm
A4 = 210 mm × 297 mm.

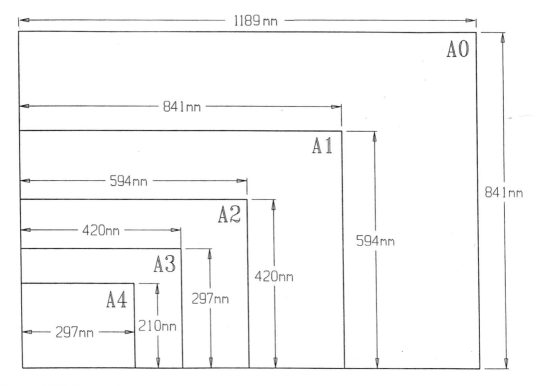

Figure 1.37 *Paper sizes*

The paper size you choose will depend upon the size of the drawing and the number of views required. Be generous, nothing looks worse than a cramped up drawing, and overcrowded dimensions. It is also false economy since overcrowding invariably leads to reading errors. As you will already have seen from some of the previous examples, the drawing should always have a border and a title block. This restricts the blank area available to draw on. Figures 1.38 and 1.39 show how the views should be positioned. These layouts are only a guide but they offer a good starting point until you become more experienced. If only one view is required then it is centred in the drawing space available.

Title block

A typical title block was shown in Figure 1.19. If you refer back to this figure you will see that it is expandable vertically and horizontally to accommodate any written information that is required. The title block should contain:

- the drawing number (which should be repeated in the top left-hand corner of the drawing)
- the drawing name (title)
- the drawing scale
- the projection used (standard symbol)
- the name and signature of the draughtsperson together with the date on which the drawing was signed
- the name and signature of the person who checks and/or approves the drawing, together with the date of signing
- the issue number and its release date
- any other information as dictated by company policy.

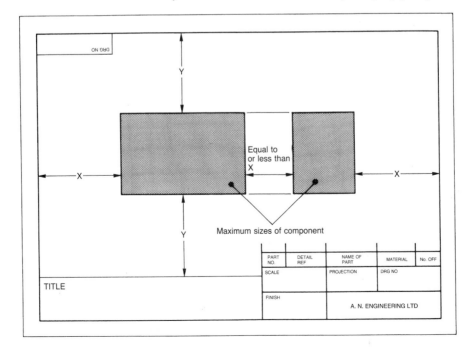

Figure 1.38 *Positioning the drawing*

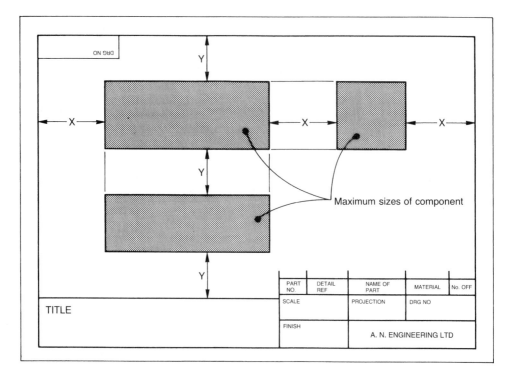

Figure 1.39 *Positioning the drawing*

Scale

The scale should be stated on the drawing as a ratio. The recommended scales are as follows:

- Full size = 1:1
- Reduced scales (smaller than full size) are:

     1:2   1:5   1:10
     1:20   1:50   1:100
     1:200 1:500 1:1000

  (NEVER use the words full-size, half-size, quarter-size, etc.).
- Enlarged scales (larger than full size) are:

     2:1   5:1   10:1
     20:1   50:1   100:1

Production of the drawing

Lines and linework

The lines of a drawing should be uniformly black, dense and bold. On any one drawing they should all be in pencil or in black ink. Pencil is quicker to use but ink prints out more clearly. Lines should be thick or thin as recommended later. Thick lines should be twice as thick as thin lines. Figure 1.40 shows the types of lines recommended in BS 308 for use in engineering drawing and how the lines should be used. This is reinforced by Table 1.1.

Sometimes the lines overlap in different views. When this happens, as shown in Figure 1.41, the following order of priority should be observed.

- Visible outlines and edges (type A) take priority over all other lines.
- Next in importance are hidden outlines and edges (type E).
- Then cutting planes (type G).
- Next come centre lines (type F and B).
- Outlines and edges of adjacent parts, etc. (type H).
- Finally, projection lines and shading lines (type B).

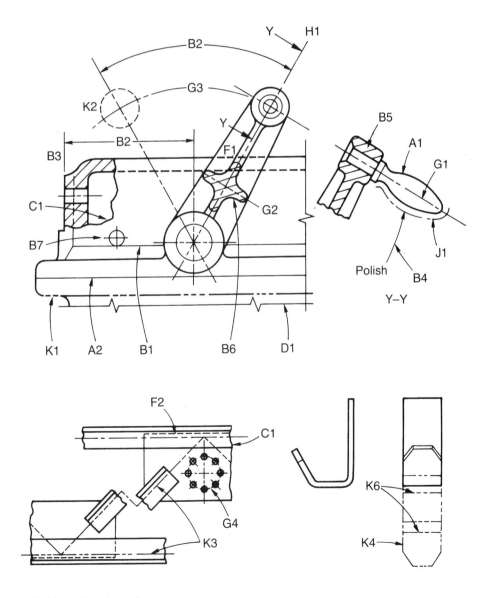

Figure 1.40 *Use of various line types*

Table 1.1 *Types of line*

Types of line		
Line	**Description**	**Application**
A	Continuous thick	A1 Visible outlines A2 Visible edges
B	Continuous thin	B1 Imaginary lines of intersection B2 Dimension lines B3 Projection lines B4 Leader lines B5 Hatching B6 Outlines of revolved sections B7 Short centre lines
C D	Continuous thin irregular Continuous thin straight with zigzags	C1 Limits of partial or interrupted views and sections, if the limit is not an axis †D1 Limits of partial or interrupted views and sections, if the limit is not an axis
E F	Dashed thick Dashed thin‡	E1 Hidden outlines E2 Hidden edges F1 Hidden outlines F2 Hidden edges
G	Chain thin	G1 Centre lines G2 Lines of symmetry G3 Trajectories and loci G4 Pitch lines and pitch circles
H	Chain thin, thick at ends and changes of direction	H1 Cutting planes
J	Chain thick	J1 Indication of lines or surfaces to which a special requirement applies (drawn adjacent to surface)
K	Chain thin double dashed	K1 Outlines and edges of adjacent parts K2 Outlines and edges of alternative and extreme positions of movable parts K3 Centroidal lines K4 Initial outlines prior to forming §K5 Parts situated in front of a cutting plane K6 Bend lines on developed blanks or patterns

NOTE. The lengths of the long dashes shown for lines G, H, J and K are not necessarily typical due to the confines of the space available.

†This type of line is suited for production of drawings by machines.

‡The thin F type line is more common in the UK, but on any one drawing or set of drawings only one type of dashed line should be used.

§Included in ISO 128-1982 and used mainly in the building industry.

Figure 1.42 shows a simple drawing using a variety of lines. List the numbers and against each number write down whether the type of line chosen is correct or incorrect. If incorrect state what type of line should have been used.

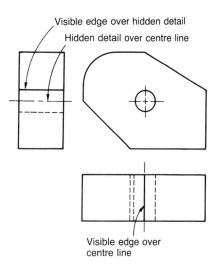

Figure 1.41 *Line priorities*

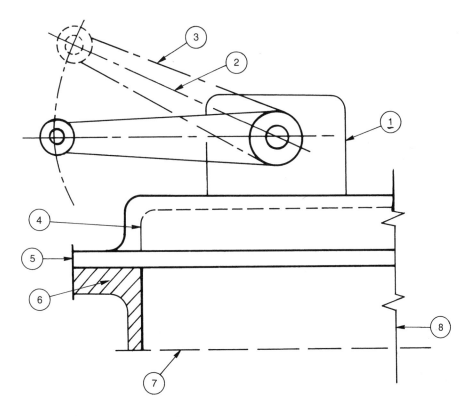

Figure 1.42 *See Test your knowledge 1.15*

Leader lines

Leader lines, as their name implies, lead written information or dimensions to the points where they apply. Leader lines are thin lines (type B) and they end in an arrowhead or in a dot, as shown in Figure 1.43(a). Arrowheads touch and stop on a line, whilst dots should always be used within an outline.

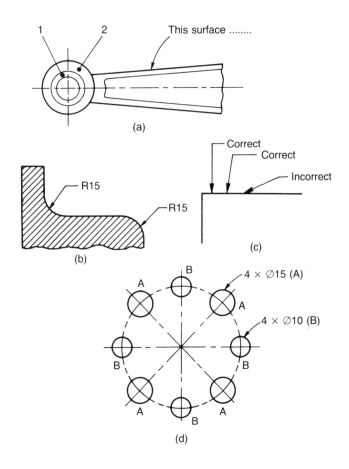

Figure 1.43 *Examples of the use of leader lines*

- When an arrowed leader line is applied to an arc it should be in line with the centre of the arc, as shown in Figure 1.43(b).
- When an arrowed leader line is applied to a flat surface, it should be nearly normal to the lines representing that surface, as shown in Figure 1.43(c).
- Long and intersecting leader lines should not be used, even if this means repeating dimensions and/or notes, as shown in Figure 1.43(d).
- Leader lines must not pass through the points where other lines intersect.
- Arrowheads should be triangular with their length some three times larger than the maximum width. They should be formed from straight lines and the arrowheads should be filled in. The arrowhead should be symmetrical about the leader line, dimension line or stem. It is recommended that arrowheads on dimension and leader lines should be some 3 – 5 mm long.
- Arrowheads showing direction of movement or direction of viewing should be some 7 – 10 mm long. The stem should be the same length as the arrowhead or slightly greater. It must never be shorter.

Test your knowledge 1.16

Figure 1.44 shows some applications of leader lines with arrowheads and leader lines with dots. List the numbers and state whether each is correct or incorrect. If incorrect explain (with sketches if required) how the drawing should be corrected.

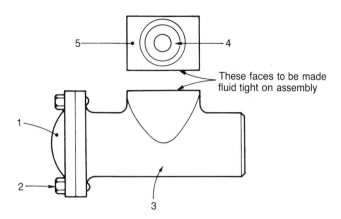

Figure 1.44 *See Test your knowledge 1.16*

Letters and numerals

Style

The style should be clear and free from embellishments. In general, capital letters should be used. A suitable style could be:

ABCDEFGHIJKLMNOPQRSTUVWXYZ
1234567890

Size

The characters used for dimensions and notes on drawings should be not less than 3 mm tall. Title and drawing numbers should be at least twice as big.

Direction of lettering

Notes and captions should be positioned so that they can be read in the same direction as the information in the title block. Dimensions have special rules and will be dealt with later.

Location of notes

General notes should all be grouped together and not scattered about the drawing. Notes relating to a specific feature should be placed adjacent to that feature.

Emphasis

Characters, words and/or notes should not be emphasized by underlining. Where emphasis is required the characters should be enlarged.

Symbols and abbreviations

If all the information on a drawing was written out in full, the drawing would become very cluttered. Therefore symbols and abbreviations are used to shorten written notes. Those recommended for use on engineering drawings are listed in BS 308, and in the corresponding student version PP7308.

With reference to BS 308 or PP7308, complete Table 1.2.

Table 1.2 *See Test your knowledge 1.17*

Abbreviation or symbol	Term
AF	
ASSY	
CRS	
	Centre line
	Countersink
	Counterbore
Ø	
DAG	
	Hexagon
	Internal
LH	
MATL	
	Maximum
	Minimum
	Number
PCD	
	Radius (in a note)
	Radius (preceding a dimension)
REQD	
RH	
SCR	
SH	
SK	
SPEC	
	Figure
	Diameter (in a note)
CYL	
CHAM	
CH HD	
	Equally spaced

Conventions

These are a form of 'shorthand' used to speed up the drawing of common features in regular use. The full range of conventions and examples of their use can be found in BS 308 or PP7308, so I will not waste space by listing them here. However by completing the next exercise you will use some of the more common conventions and this will help you to become familiar with them.

With reference to BS 308 or
PP7308, complete Figure 1.45.
Note that you must take care to
use the same types of lines as
shown in the standard or the
conventions will become
meaningless. This applies
particularly to line thickness.

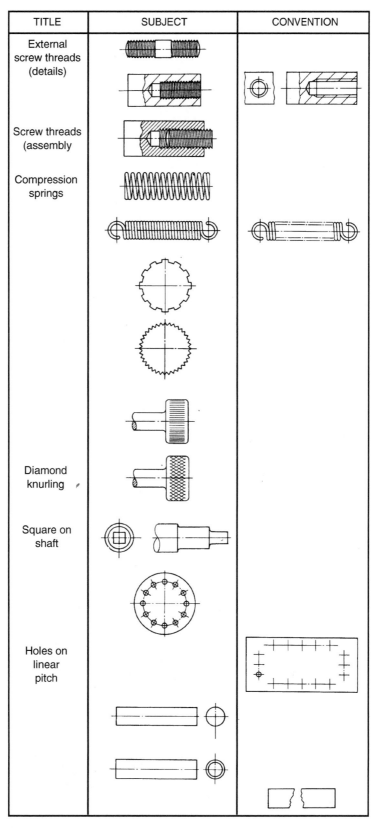

TITLE	SUBJECT	CONVENTION
External screw threads (details)		
Screw threads (assembly		
Compression springs		
Diamond knurling		
Square on shaft		
Holes on linear pitch		

Figure 1.45 *See Test your knowledge 1.18*

Dimensioning

When a component is being dimensioned, the dimension lines and the projection lines should be thin full lines (type B). Where possible dimensions should be placed outside the outline of the object, as shown in Figure 1.46(a). The rules are:

- Outline of object to be dimensioned in thick lines (type A).
- Dimension and projection lines should be half the thickness of the outline (type B).
- There should be a small gap between the projection line and the outline.
- The projection line should extend to just beyond the dimension line.
- Dimension lines end in an arrowhead that should touch the projection line to which it refers.
- All dimensions should be placed in such a way that they can be read from the bottom right-hand corner of the drawing.

The purpose of these rules is to allow the outline of the object to stand out prominently from all the other lines and to prevent confusion.

There are three ways in which a component can be dimensioned. These are:

- Chain dimensioning, as shown in Figure 1.46(b).
- Absolute dimensioning (dimensioning from a datum) using parallel dimension lines, as shown in Figure 1.46(c).
- Absolute dimensioning (dimensioning from a datum using superimposed running dimensions, as shown in Figure 1.46(d)). Note the common origin (termination) symbol.

It is neither possible to manufacture an object to an exact size nor to measure an exact size. Therefore important dimensions have to be toleranced. That is, the dimension is given two sizes: an upper limit of size and a lower limit of size. Providing the component is made so that it lies between these limits it will function correctly. Information on Limits and Fits can be found in BS 4500.

The method of dimensioning can also affect the accuracy of a component and produce some unexpected effects. Figure 1.46(b) shows the effect of chain dimensioning on a series of holes or other features.

The designer specifies a common tolerance of ±0.2 mm. However, since this tolerance is applied to each and every dimension, the cumulative tolerance becomes ±0.6 mm by the time you reach the final, right-hand hole, which is not what was intended. Therefore, absolute dimensioning as shown in Figure 1.46(c) and (d) is to be preferred in this example.

With absolute dimensioning, the position of each hole lies within a tolerance of ±0.2 mm and there is no cumulative error. Further examples of dimensioning techniques are shown in Figure 1.47.

It is sometimes necessary to indicate machining processes and surface finish. The machining symbol, together with examples of process notes and the surface finishes in micrometres (μm), is shown in Figure 1.48.

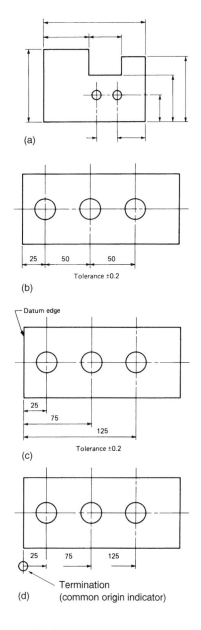

Figure 1.46 *Dimensioning*

Sectioning

Sectioning is used to show the hidden detail inside hollow objects more clearly than can be achieved using dashed thin (type E) lines. Figure 1.50(a) shows an example of a simple sectioned drawing. The cutting plane is the line A–A. In your imagination you remove everything to the left of the cutting plane, so that you only see what remains to the right of the cutting plane looking in the direction of the arrowheads. Another example is shown in Figure 1.50(b).

Figure 1.50(c) shows how to section an assembly. Note how solid shafts and the key are not sectioned. Also note that thin webs that lie on the section plane are not sectioned.

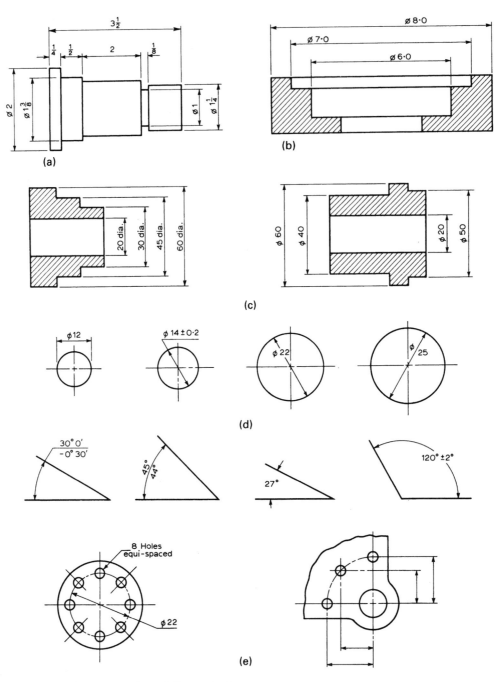

Figure 1.47 *More examples of dimensioning*

When interpreting sectioned drawings, some care is required. It is easy to confuse the terms Sectional view and Section.

Sectional view

In a sectional view you see the outline of the object at the cutting plane. You also see all the visible outlines seen beyond the cutting plane in the direction of viewing. Therefore, Figure 1.50(a) is a sectional view.

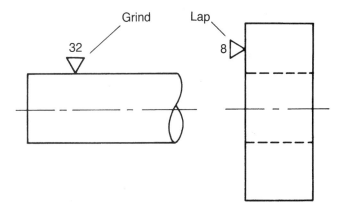

Figure 1.48 *Indicating surface finishes*

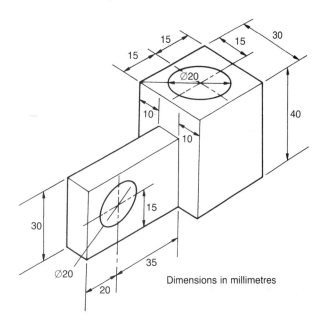

Dimensions in millimetres

Figure 1.49 *See Test your knowledge 1.19*

Test your knowledge 1.19

Figure 1.49 shows a component drawn in isometric projection. Redraw it in first-angle orthographic projection and add the dimensions, using the following techniques:

(a) Absolute dimensioning using parallel dimension lines.
(b) Absolute dimensioning using superimposed running dimensions.

Test your knowledge 1.20

(a) Redraw Figure 1.50(a) as a section (remember that the drawing is a sectional view).
(b) Explain why Figure 1.50(b) can be a section or a sectional view.

Section

A section only shows the outline of the object at the cutting plane. Visible outlines beyond the cutting plane in the direction of viewing are not shown. Therefore a section has no thickness.

Cutting planes

You have already been introduced to cutting planes in the previous examples. They consist of type G lines; that is, a thin chain line that is thick at the ends and at changes of direction. The direction of viewing is shown by arrows with large heads. The points of the arrowheads touch the thick portion of the cutting plane. The cutting plane is labelled by placing a capital letter close to the stems of the arrows. The same letters are used to identify the corresponding section or sectional view.

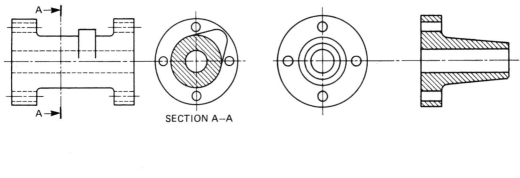

SECTION A–A

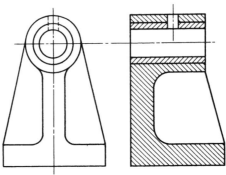

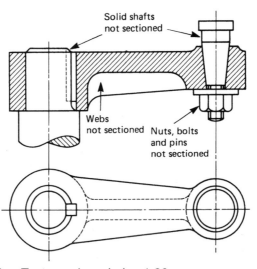

Figure 1.50 (a) – (c)   *See Test your knowledge 1.20*

Hatching

You will have noticed that the shading of sections and sectional views consists of sloping, thin (type B) lines. This is called hatching. The lines are equally spaced, slope at 45° and are not usually less than 4 mm apart. However when hatching very small areas the hatching can be reduced, but never less than 1 mm. The drawings in this book may look as though they do not obey these rules. Remember that they have been reduced from much bigger drawings to fit onto the pages.

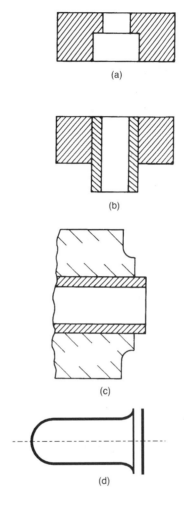

Figure 1.51 *Rules of hatching*

Figure 1.51 shows the basic rules of hatching. The hatching of separated areas is shown in Figure 1.51(a). Separate sectioned areas of the same component should be hatched in the same direction and with the same spacing.

Figure 1.51(b) shows how to hatch assembled parts. Where the different parts meet on assembly drawings, the direction of hatching should be reversed. The hatching lines should also be staggered. The spacing may also be changed.

Figure 1.51(c) shows how to hatch large areas. This saves time and avoids clutter. The hatching is limited to that part of the area that touches adjacent hatched parts or just to the outline of a large part.

Figure 1.51(d) shows how sections through thin materials can be blocked in solid rather than hatched. There should be a gap of not less than 1 mm between adjacent parts even when these are a tight fit in practice.

Finally I have included some further examples of sectioning in Figure 1.52. These include assemblies, half sections, part sections and revolved sections. Then it will be your turn to produce some engineering drawings including some or all of the features outlined in this section.

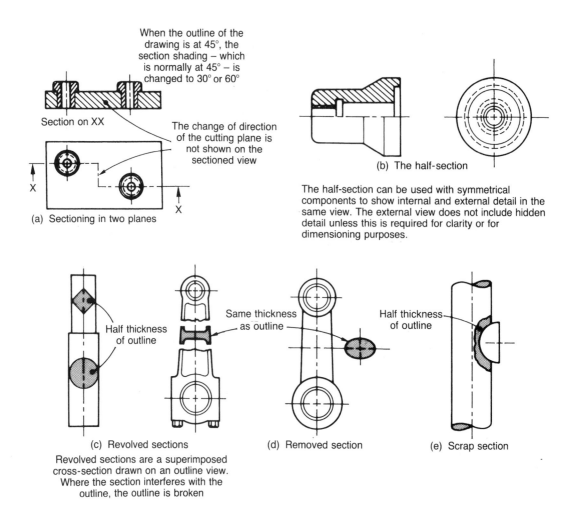

When the outline of the drawing is at 45°, the section shading – which is normally at 45° – is changed to 30° or 60°

Section on XX

The change of direction of the cutting plane is not shown on the sectioned view

(a) Sectioning in two planes

(b) The half-section

The half-section can be used with symmetrical components to show internal and external detail in the same view. The external view does not include hidden detail unless this is required for clarity or for dimensioning purposes.

Half thickness of outline

Same thickness as outline

Half thickness of outline

(c) Revolved sections

Revolved sections are a superimposed cross-section drawn on an outline view. Where the section interferes with the outline, the outline is broken

(d) Removed section

(e) Scrap section

Figure 1.52 *Examples of sectioning*

Activity 1.8

(a) Redraw Figure 1.53 in third-angle projection. Include an end-view looking in the direction of arrow A, and section the elevation on the cutting plan XX.

(b) Figure 1.54 shows a cast iron bend.
(i) Redraw, adding an end view looking in the direction of arrow A.
(ii) Section the elevation on the centre line.
(iii) Draw an auxiliary view of flange B.

Schematic circuit diagrams

I introduced you to examples of electrical, pneumatic and hydraulic schematic circuit diagrams earlier in this unit. We are now going to look at the drawing of such circuits in greater detail. In all these circuits, the circuit components are not drawn out in detail but are represented by symbols. To read the circuit diagram you must understand what the symbols stand for and how their function is interpreted.

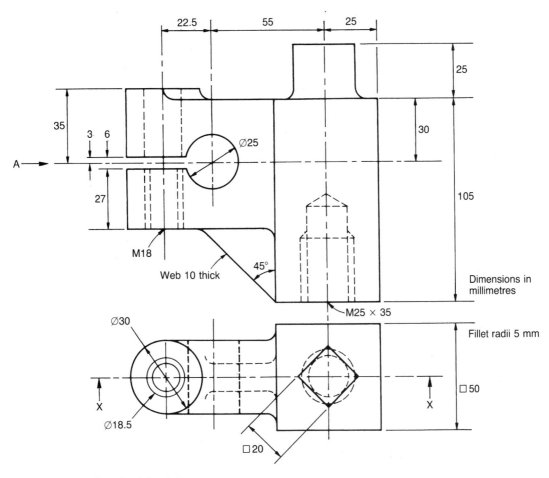

Figure 1.53 *See Activity 1.8*

As mentioned earlier, you will need access to BSI publication PP7307: Graphical Symbols for Use in Colleges and Schools. It summarizes the basic essentials of a number of full standard specifications relating to schematic circuit diagrams and their related wiring and piping diagrams.

Fluid power schematic diagrams

These diagrams cover both pneumatic and hydraulic circuits. The symbols that we shall use do not illustrate the physical make-up, construction or shape of the components. Neither are the symbols to scale or orientated in any particular position. They are only intended to show the 'function' of the component they portray, the connections and the fluid flow path.

Complete symbols are made up from one or more basic symbols and from one or more functional symbols. Examples of some basic symbols are shown in Figure 1.55 and some functional symbols are shown in Figure 1.56.

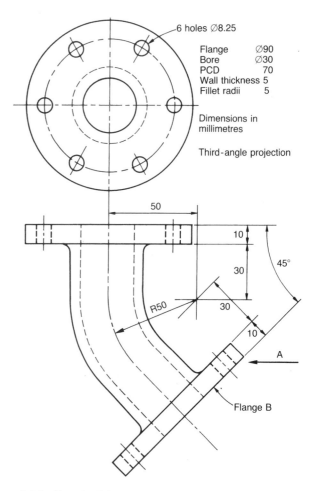

Figure 1.54 *See Activity 1.8*

Energy converters

Let's now see how we can combine some of these basic and functional symbols to produce a complete symbol representing a component. For example let's start with a motor. The complete symbol is shown in Figure 1.57.

The large circle indicates that we have an energy conversion unit such as a motor or pump. Notice that the fluid flow is into the device and that it is pneumatic. The direction of the arrowhead indicates the direction of flow. The fact that the arrowhead is clear (open) indicates that the fluid is air. Therefore the device must be a motor. If it was a pump the fluid flow would be out of the circle. The single line at the bottom of the circle is the outlet (exhaust) from the motor and the double line is the mechanical output from the motor.

Now let's analyse the symbol shown in Figure 1.58.

- The circle tells us that it is an energy conversion unit.
- The arrowheads show that the flow is from the unit so it must be a pump.

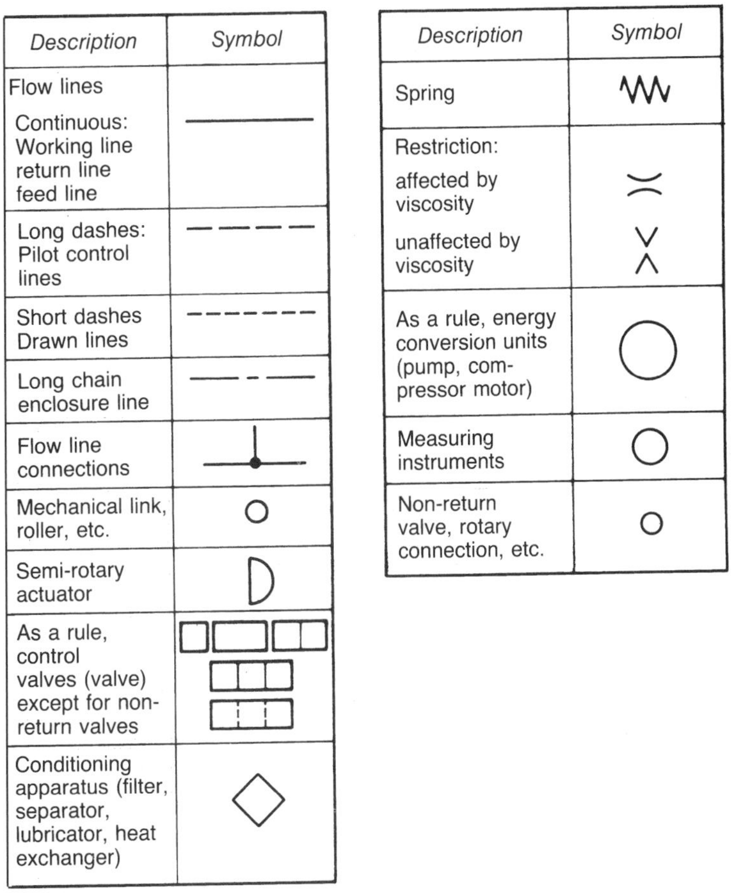
Description
Symbol
Flow lines
Continuous: Working line return line feed line
Long dashes: Pilot control lines
Short dashes Drawn lines
Long chain enclosure line
Flow line connections
Mechanical link, roller, etc.
Semi-rotary actuator
As a rule, control valves (valve) except for non-return valves
Conditioning apparatus (filter, separator, lubricator, heat exchanger)
Description
Symbol
Spring
Restriction:
affected by viscosity
unaffected by viscosity
As a rule, energy conversion units (pump, compressor motor)
Measuring instruments
Non-return valve, rotary connection, etc.

Figure 1.55 *Basic symbols used in fluid power diagrams*

Test your knowledge 1.21

(a) Draw the symbol for a unidirectional, fixed displacement pneumatic pump (compressor).
(b) Draw the symbol for a fixed capacity hydraulic motor.

- The arrowheads are solid so it must be a hydraulic pump.
- The arrowheads point in opposite directions so the pump can deliver the hydraulic fluid in either direction depending upon its direction of rotation.
- The arrow slanting across the pump is the variability symbol, so the pump has variable displacement.
- The double lines indicate the mechanical input to the pump from some engine or motor.

Summing up, we have a variable displacement, hydraulic pump that is bi-directional.

Directional control valves

The function of a directional control valve is to open or close flow lines in a system. Control valve symbols are always drawn in square boxes or groups of square boxes to form a rectangle. This is how you recognize them. Each box indicates a discrete position for the control valve. Flow paths through a valve are known as 'ways'. Thus a 4-way valve has four flow paths through the valve. This will be the same as the number of connections. We can, therefore, use a number

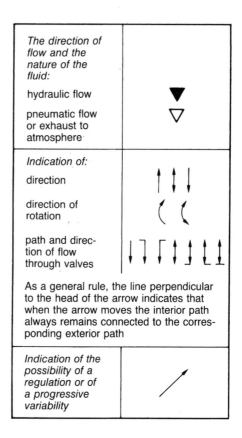

The direction of flow and the nature of the fluid: hydraulic flow pneumatic flow or exhaust to atmosphere	
Indication of: direction direction of rotation path and direction of flow through valves As a general rule, the line perpendicular to the head of the arrow indicates that when the arrow moves the interior path always remains connected to the corresponding exterior path	
Indication of the possibility of a regulation or of a progressive variability	

Figure 1.56 *Functional symbols used in fluid power diagrams*

Figure 1.57 *Basic symbol for a motor*

Figure 1.58 *Energy converter symbol (see text)*

code to describe the function of a valve. Figure 1.59 shows a 4/2 directional control valve. This valve has four flow paths, ports or connections and two positions. The two boxes indicate the two positions. The appropriate box is shunted from side to side so that, in your imagination, the internal flow paths line up with the connections. Connections are shown by the lines that extend

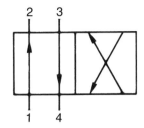

Figure 1.59 *4/2 directional control valve*

(a) State the numerical code that describes the valve shown in Figure 1.60.
(b) Describe the flow path drawn.
(c) Sketch and describe the flow path when the valve is in its alternative position.

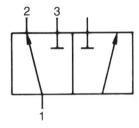

Figure 1.60 *See Test your knowledge 1.22*

'outside' the perimeters of the boxes.

As drawn, the fluid can flow into port 1 and out of port 2. Fluid can also flow into port 3 and out of port 4. In the second position, The fluid flows into port 3 and out of port 1. Fluid can also flow into port 4 and out of port 2.

Valve control methods

Before we look at other examples of directional control valves, let's see how we can control the positions of a valve. There are four basic methods of control, these are:

- Manual control of the valve position.
- Mechanical control of the valve position.
- Electromagnetic control of the valve position.
- Pressure control of the valve positions (direct and indirect).
- Combined control methods.

The methods of control are shown in Figure 1.61. With simple electrical or pressure control, it is only possible to move the valve to one, two or three discrete positions. The valve spool may be located in such positions by a spring loaded detent.

Combinations of the above control methods are possible. For example a single solenoid with spring return for a two position valve. Let's now look at some further directional control valves (DCVs).

- Figure 1.62(a) shows a 4/2 DCV controlled by a single solenoid with a spring return.
- Figure 1.62(b) shows a 4/3 DCV. That is, a directional control valve with 4 ports (connections) and 3 positions. It is operated manually by a lever with spring return to the centre. The service ports are isolated in the centre position. An application of this

Description	Symbol
Manual control : general symbol	
by push-button	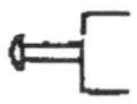
by lever	
by pedal	
Mechanical control: by plunger or tracer	
by spring	
by roller	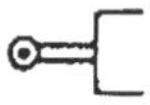
by roller, operating in one direction only	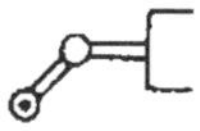
Electrical control: by solenoid (one winding)	
(two windings operating in opposite directions)	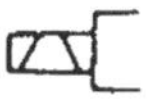
by electric motor	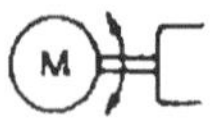

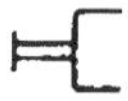

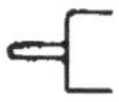

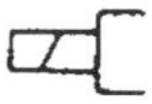

Description	Symbol
Control by application or release of pressure Direct acting control: by application of pressure	
by release of pressure	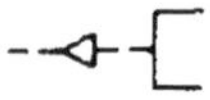
by different control areas	
Indirect control, pilot actuated: by application of pressure	
by release of pressure	
Interior control paths (paths are inside the unit)	
Combined control: by solenoid and pilot directional valve (pilot directional valve is actuated by the solenoid)	
by solenoid or pilot direction valve (either may actuate the control independently)	

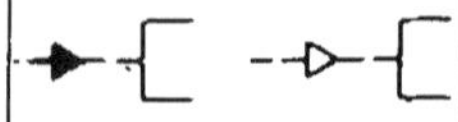

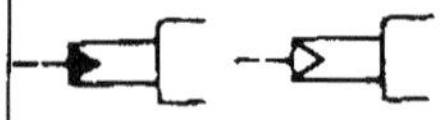

Figure 1.61 *Methods of control*

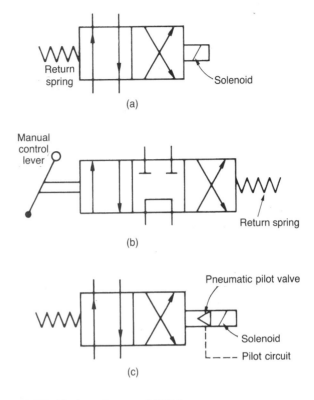

Figure 1.62 *Various types of DCV*

valve will be shown later.
- Figure 1.62(c) shows a 4/2 DCV controlled by pneumatic pressure by means of a pilot valve. The pilot valve is actuated by a single solenoid and a return spring.

Linear actuators

A linear actuator is a device for converting fluid pressure into a mechanical force capable of doing useful work and combining this force with limited linear movement. Put more simply, a piston in a cylinder. The symbols for linear actuators (also known as 'jacks' and 'rams') are simple to understand and some examples are shown in Figure 1.64.

- Figure 1.64(a) shows a single-ended, double-acting actuator. That is, the piston is connected by a piston rod to some external mechanism through one end of the cylinder only. It is double

Test your knowledge 1.23

Describe the DCV whose symbol is shown in Figure 1.63.

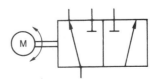

Figure 1.63 *See Test your knowledge 1.23*

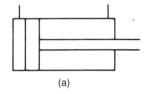

(a)

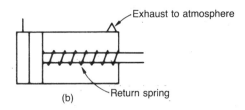

Exhaust to atmosphere

Return spring

(b)

Cushion

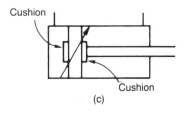

Cushion

(c)

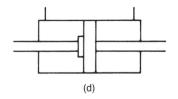

(d)

Figure 1.64 *Various types of linear actuator*

acting because fluid pressure can be applied to either side of the piston.

• Figure 1.64(b) shows a single-ended, single-acting actuator with spring return. Here the fluid pressure is applied only to one side of the piston. Note the pneumatic exhaust to atmosphere so that the air behind the piston will not cause a fluid lock.

• Figure 1.64(c) shows a single-ended, single-acting actuator, with double variable cushion damping. The cushion damping prevents the piston impacting on the ends of the cylinder and causing damage.

• Figure 1.64(d) shows a double ended, double-acting actuator fitted with single, fixed cushion damping.

We are now in a position to use the previous component symbols to produce some simple fluid power circuits.

Figure 1.65 shows a single-ended, double-acting actuator controlled by a 4/3 tandem centre, manually operated DCV. Note that in the neutral position both sides of the actuator piston are blocked off, forming a hydraulic lock. In this position the pump flow is being returned directly to the tank. Note the tank symbol. This

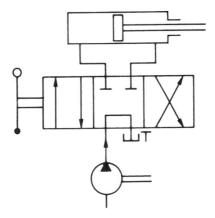

Figure 1.65 *Actuator controlled by a DCV*

system is being supplied by a single direction fixed displacement hydraulic pump.

Figure 1.66 shows a simple pneumatic hoist capable of raising a load. The circuit uses two 2-port manually operated push-button valves connected to a single-ended, single-acting actuator. Supply pressure is indicated by the circular symbol with a black dot in its centre. Valve 'b' has a threaded exhaust port indicated by the extended arrow. When valve 'a' is operated, compressed air from the air line is admitted to the underside of the piston in the cylinder. This causes the piston to rise and to raise the load. Any air above the piston is exhausted to the atmosphere through the threaded exhaust port at the top of the cylinder. Again this is indicated by a long arrow. When valve 'b' is operated, it connects the cylinder to the exhaust and the actuator is vented to the atmosphere. The load is lowered by gravity.

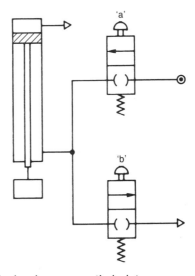

Figure 1.66 *A simple pneumatic hoist*

Both these circuits are functional, but they do not have protection against over-pressurization, neither do they have any other safety devices fitted. Therefore, we need to increase our vocabulary of components before we can design a safe, practical circuit. We will now consider the function and use of pressure and flow control valves.

Pressure relief and sequence valves

Figure 1.67 shows an example of a pressure relief (safety) valve. In Figure 1.67(a) the valve is being used in a hydraulic circuit. Pressure is controlled by opening the exhaust port to the reservoir tank against an opposing force such as a spring. In Figure 1.67(b) the valve is being used in a pneumatic circuit so it exhausts to the atmosphere.

Figures 1.67(c) and 1.67(d) show the same valves except that this time the relief pressure is variable, as indicated by the arrow drawn across the spring. If the relief valve setting is used to control the normal system pressure as well as acting as an emergency safety valve, the adjustment mechanism for the valve must be designed so

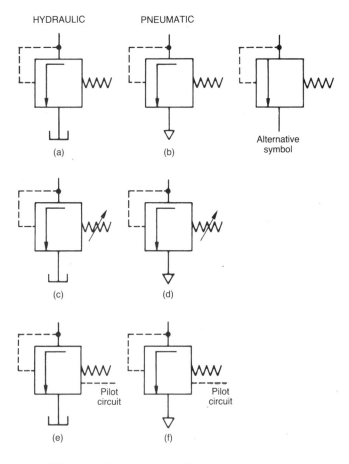

Figure 1.67 *Use of a pressure relief valve*

that the maximum safe working pressure for the circuit cannot be exceeded.

Figures 1.67(e) and 1.67(f) show the same valves with the addition of pilot control. This time the pressure at the inlet port is not only limited by the spring but also by the pressure of the pilot circuit superimposed on the spring. The spring offers a minimum pressure setting and this can be increased by increasing the pilot circuit pressure up to some pre-determined safe maximum. Sometimes the spring is omitted and only pilot pressure is used to control the valve.

Sequence valves are closely related to relief valves in both design and function and are represented by very similar symbols. They permit the hydraulic fluid to flow into a sub-circuit, instead of back to the reservoir, when the main circuit pressure reaches the setting of the sequence valve. You can see that Figure 1.68 is very similar to a pressure relief valve (PRV) except that, when it opens, the fluid is directed to the next circuit in the sequence instead of being exhausted to the reservoir tank or allowed to escape to the atmosphere.

Flow control valves

Flow control valves, as their name implies, are used in systems to control the rate of flow of fluid from one part of the system to another. The simplest valve is merely a fixed restrictor. For operational reasons this type of flow control valve is inefficient, so the restriction is made variable as shown in Figure 1.69(a). This is a throttling valve. The full symbol is shown in Figure 1.69(b). In this example the valve setting is being adjusted mechanically. The valve rod ends in a roller follower in contact with a cam plate.

Sometimes it is necessary to ensure that the variation in inlet pressure to the valve does not affect the flow rate from the valve. Under these circumstances we use a pressure compensated flow control valve (PCFCV). The symbol for this type of valve is shown in Figure 1.70. This symbol suggests that the valve is a combination of a variable restrictor and a pilot operated relief valve. The enclosing box is drawn using a long-chain line. This signifies that the components making up the valve are assembled as a single unit.

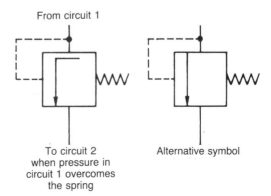

Figure 1.68 *Sequence valve*

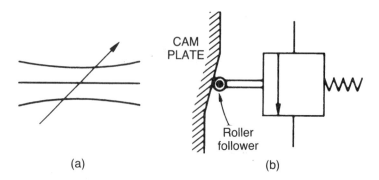

Figure 1.69 *Fluid control valves*

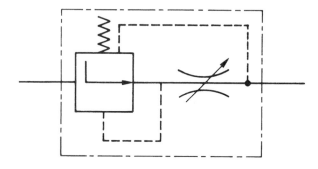

Figure 1.70 *Pressure compensated flow control valve*

Non-return valves and shuttle valves

The non-return valve (NRV), or check valve as it is sometimes known, is a special type of directional control valve. It only allows the fluid to flow in one direction and it blocks the flow in the reverse direction. These valves may be operated directly or by a pilot circuit. Some examples are shown in Figure 1.71.

- Figure 1.71(a) shows a valve that opens (is free) when the inlet pressure is higher than the outlet pressure (back-pressure).
- Figure 1.71(b) shows a spring-loaded valve that only opens when the inlet pressure can overcome the combined effects of the outlet pressure and the force exerted by the spring.
- Figure 1.71(c) shows a pilot controlled NRV. It only opens if the inlet pressure is greater than the outlet pressure. However, these pressures can be augmented by the pilot circuit pressure.
  (i) The pilot pressure is applied to the inlet side of the NRV. We now have the combined pressures of the main (primary) circuit and the pilot circuit acting against the outlet pressure. This enables the valve to open at a lower main circuit pressure than would normally be possible.
  (ii) The pilot pressure is applied to the outlet side of the NRV. This assists the outlet or back-pressure in holding the valve closed. Therefore it requires a greater main circuit pressure to open the valve. By adjusting the pilot pressure in these two examples we can control the circumstances under which the NRV opens.

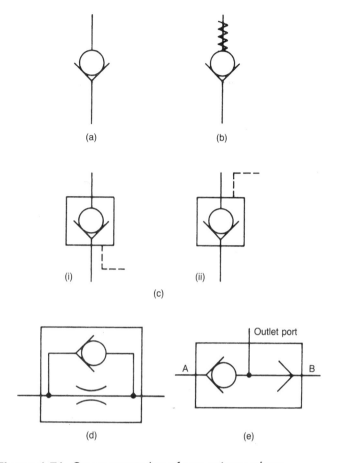

Figure 1.71 *Some examples of non-return valves*

- Figure 1.71(d) shows a valve that allows normal full flow in the forward direction but restricted flow in the reverse direction. The valves previously discussed did not allow any flow in the reverse direction.
- Figure 1.71(e) shows a simple shuttle valve. As its name implies, the valve is able to shuttle backwards and forwards. There are two inlet ports and one outlet port. Imagine that inlet port A has the higher pressure. This pressure overcomes the inlet pressure at B and moves the shuttle valve to the right. The valve closes inlet port B and connects inlet port A to the outlet port. If the pressure at inlet port B rises, or that at A falls, the shuttle will move back to the left. This will close inlet port A and connect inlet port B to the outlet. Thus, the inlet port with the higher pressure is automatically connected to the outlet port.

Conditioning equipment

The working fluid, be it oil or air, has to operate in a variety of environments and it can become overheated and/or contaminated. As its name implies, conditioning equipment is used to maintain the fluid in its most efficient operating condition. A selection of conditioning equipment symbols are shown in Figure 1.72. Note that all conditioning device symbols are diamond shaped.

Filters and *strainers* have the same symbol. They are normally identified within the system by their position. The filter element (dotted line) is always positioned at 90° to the fluid path.

Water traps are easily distinguished from filters since they have a drain connection and an indication of trapped water. Water traps are particularly important in pneumatic systems because of the humidity of the air being compressed.

Lubricators are particularly important in pneumatic systems. Hydraulic systems using oil are self-lubricating. Pneumatic systems use air, which has no lubricating properties so oil, in the form of a mist, has to be added to the compressed air line.

Heat exchangers can be either heaters or coolers. If the hydraulic oil becomes too cool it becomes thicker (more viscous) and the system becomes sluggish. If the oil becomes too hot it will become too thin (less viscous) and not function properly. The direction of the arrows in the symbol indicates whether heat energy is taken from the fluid (cooler) or given to the fluid (heater). Notice that the cooler can show the flow lines of the coolant.

Filters, water traps, lubricators and miscellaneous apparatus

Description	Symbol
Filter or strainer	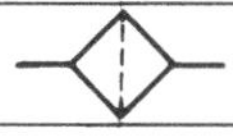
Water trap: with manual control	
automatically drained	
Filter with water trap: with manual control	
automatically drained	
Air dryer	
Lubricator	
Conditioning unit detailed symbol simplified symbol	

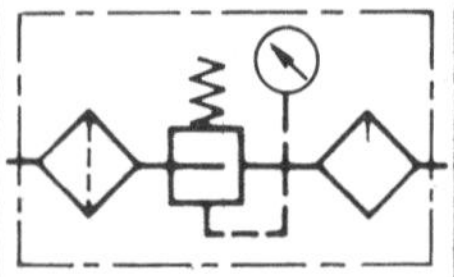

Heat exchangers

Description	Symbol
Temperature controller (arrows indicate that heat may be either introduced or dissipated)	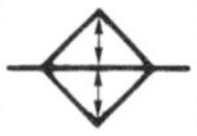
Cooler (arrows indicate the extraction of heat) without representation of the flow lines of the coolant	
with representation of the flow lines of the coolant	
Heater (arrows indicate the introduction of heat)	

Figure 1.72 *Symbols for conditioning devices*

There is one final matter to be considered before you can try your hand at designing a circuit, and that is the pipework circuit to connect the various components together. The correct way of representing pipelines is shown in Figure 1.74.

- Figure 1.74(a) shows pipelines that are crossing each other but are not connected.
- Figure 1.74(b) shows three pipes connected at a junction. The junction (connection) is indicated by the solid circle (or large dot, if you prefer).
- Figure 1.74(c) shows four pipes connected at a junction. On no account can the connection be drawn as shown in Figure 1.74(d). This is because there is always a chance of the ink running where lines cross on a drawing. The resulting 'blob' could then be misinterpreted as a connection symbol with disastrous results.

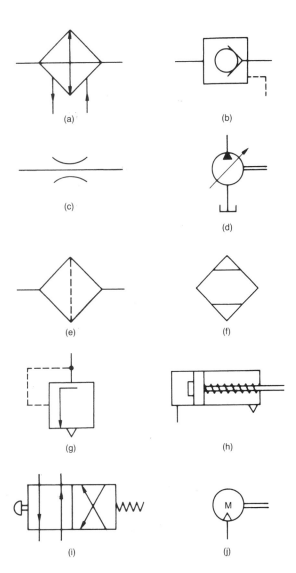

Figure 1.73 *See Test your knowledge 1.24*

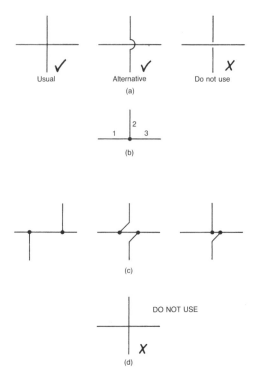

Figure 1.74 *Representing pipelines*

Activity 1.9

Figure 1.75 shows the general principles for the hydraulic drive to the ram of a shaping machine. The ram is moved backwards and forwards by a double-acting single-ended hydraulic actuator. The drawing was made many years ago and it uses outdated symbols. Using current symbols and practices (as set out in BS PP7307) you are to draw a schematic hydraulic circuit diagram for the machine.

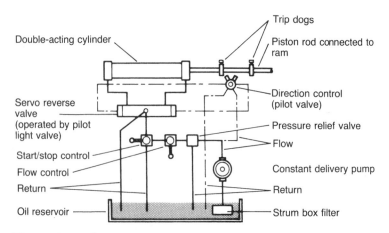

Figure 1.75 *See Activity 1.9*

Electronic circuits

Electrical and electronic circuits can also be drawn using schematic symbols to represent the various components. The full range of symbols and their usage can be found in BS 3939. This is a very extensive standard and well beyond the needs of this book. For our immediate requirements you should refer to PP7307 Graphical Symbols for Use in Schools and Colleges. Figure 1.76 shows a selection of symbols that will be used in the following examples.

- A *cell* is a source of direct current (d.c.) electrical energy. Primary cells have a nominal potential of 1.5 volts each. They cannot be recharged and are disposable. Secondary cells are rechargeable. Lead–acid cells have a nominal potential of 2 volts and nickel cadmium (NiCd) cells have a nominal potential of 1.2 volts. Cells are often connected in series to form a battery.
- *Batteries* consist of a number of cells connected in series to increase the overall potential. A 12 volt car battery consists of six lead–acid secondary cells of 2 volts each.
- *Fuses* protect the circuit in which they are connected from excess current flow. This can result from a fault in the circuit, from a fault in an appliance connected to the circuit or from too many appliances being connected to the same circuit. The current flowing in the circuit tends to heat up the fuse wire. When the current reaches some pre-determined value the fuse wire melts and breaks the circuit so the current can no longer flow. Without a fuse the circuit wiring could overheat and cause a fire.
- *Resistors* are used to control the magnitude of the current flowing in a circuit. The resistance value of the resistor may be fixed or it may be variable. Variable resistors may be pre-set or they may be adjustable by the user. The electric current does work in flowing through the resistor and this heats up the resistor. The resistor must be chosen so that it can withstand this heating effect and sited so that it has adequate ventilation.
- *Capacitors*, like resistors, may be fixed in value or they may be preset or variable. Capacitors store electrical energy but, unlike secondary cells, they may be charged or discharged almost instantaneously. The stored charge is much smaller than the charge stored by a secondary cell. Large value capacitors are used to smooth the residual ripple from the rectifier in a power pack. Medium value capacitors are used for coupling and decoupling the stages of audio frequency amplifiers. Small value capacitors are used for coupling and decoupling radio frequency signals and they are also used in tuned (resonant) circuits.
- *Inductors* act like electrical 'flywheels'. They limit the build up of current in a circuit and try to keep the circuit running by putting energy back into it when the supply is turned off. They are used as current limiting devices in fluorescent lamp units. They are used as chokes in telecommunications equipment. They are also used together with capacitors to make up resonant (tuned) circuits in telecommunications equipment.
- *Transformers* are used to raise or lower the voltage of alternating currents. Inductors and transformers cannot be used in direct current circuits. You cannot get something for nothing, so if you

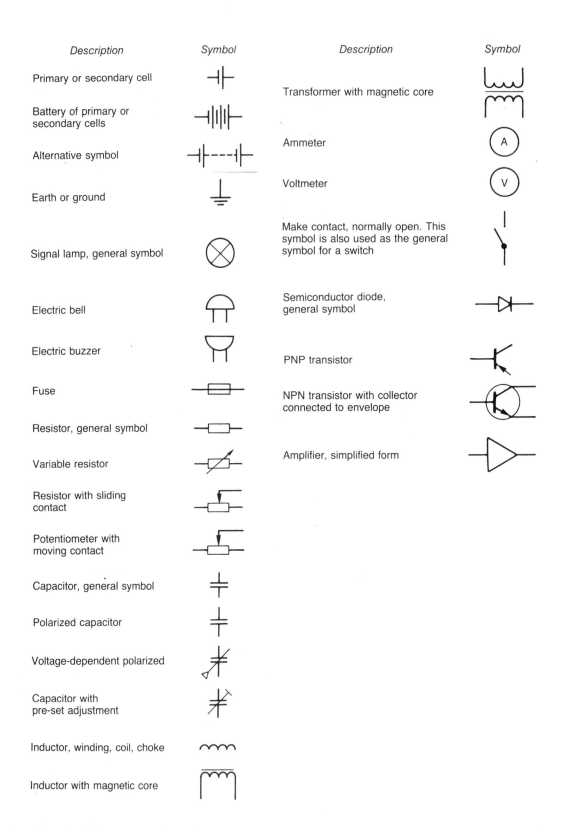

Description	Symbol	Description	Symbol
Primary or secondary cell		Transformer with magnetic core	
Battery of primary or secondary cells			
Alternative symbol		Ammeter	
Earth or ground		Voltmeter	
Signal lamp, general symbol		Make contact, normally open. This symbol is also used as the general symbol for a switch	
Electric bell		Semiconductor diode, general symbol	
Electric buzzer		PNP transistor	
Fuse		NPN transistor with collector connected to envelope	
Resistor, general symbol			
Variable resistor		Amplifier, simplified form	
Resistor with sliding contact			
Potentiometer with moving contact			
Capacitor, general symbol			
Polarized capacitor			
Voltage-dependent polarized			
Capacitor with pre-set adjustment			
Inductor, winding, coil, choke			
Inductor with magnetic core			

Figure 1.76 *Electronic symbols*

increase the voltage you decrease the current accordingly so that (neglecting losses), $V \times I = k$ where k is a constant for the primary and secondary circuits of any given transformer.

- *Ammeters* measure the current flowing in a circuit. They are always wired in series with the circuit so that the current being measured can flow through the meter.
- *Voltmeters* measure the potential difference (voltage) between two points in a circuit. To do this they are always wired in parallel across that part of the circuit where the potential is to be measured.
- *Switches* are used to control the flow of current in a circuit. They can only open or close the circuit. So the current either flows or it doesn't.
- *Diodes* are like the non-return valves in hydraulic circuits. They allow the current to flow in one direction only as indicated by the arrowhead of the symbol. They are used to rectify alternating current (a.c.) and convert it into d.c.
- *Transistors* are used in high-speed switching circuits and to magnify radio and audio frequency signals.
- *Integrated circuits* consist of all the components necessary to produce amplifiers, oscillators, central processor units, computer memories and a host of other devices fabricated onto a single slice of silicon; each chip being housed in a single compact package.

Let's look at some examples of schematic circuit diagrams using these symbols. All electric circuits consist of:

- A source of electrical energy (e.g. a battery or a generator).
- A means of controlling the flow of electric current (e.g. a switch or a variable resistor).
- An appliance to convert the electrical energy into useful work (e.g. a heater, a lamp, or a motor).
- Except for low-power battery operated circuits, an over-current protection device (fuse or circuit breaker).
- Conductors (wires) to connect these various circuit elements together. Note that the rules for drawing conductors that are connected and conductors that are crossing but not connected are the same as for drawing pipework as previously described in Figure 1.74.

Figure 1.77 shows a very simple circuit that satisfies the above requirements. In Figure 1.77(a) the switch is 'closed' therefore the circuit as a whole is also a closed loop. This enables the electrons that make up the electric current to flow from the source of electrical energy through the appliance (lamp) and back to the source of energy ready to circulate again. Rather like the fluid in our earlier hydraulic circuits. In Figure 1.77(b) the switch is 'open' and the circuit is no longer a closed loop. The circuit is broken. The electrons can no longer circulate. The circuit ceases to function. We normally draw our circuits with the switches in the 'open' position so that the circuit is not functioning and is 'safe'.

Figure 1.78 shows a simple battery operated circuit for determining the resistance of a fixed value resistor. The resistance value is obtained by substituting the values of current and potential into the formula, $R = V/I$. The current in amperes is read from the ammeter and the potential in volts is read from the voltmeter. Note that the ammeter is wired in series with the resistor so that the current can flow through it. The voltmeter is wired in parallel with the resistor so that the potential can be read across it. This is always the way these instruments are connected.

Figure 1.79 shows a circuit for operating the light over the stairs in a house. The light can be operated either by the switch at the bottom of the stairs or by the switch at the top of the stairs. Can you work out how this is achieved? The switches are of a type called 'two-way, single-pole'. The circuit is connected to the mains supply. It is protected by a fuse in the 'consumer unit'. This unit contains the main switch and all the fuses for the house and is situated adjacent to the supply company's meter and main fuse.

Figure 1.80 shows a two-stage transistorized amplifier. It also shows a suitable power supply. Table 1.3 lists and names the components.

Figure 1.81 shows a similar amplifier using a single chip. Such an amplifier would have the same performance but fewer components are required. Therefore it is cheaper and quicker to make.

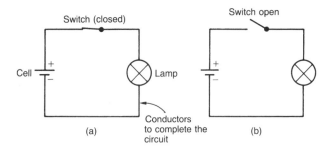

Figure 1.77 *A simple electric circuit*

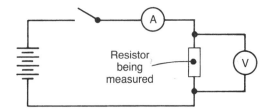

Figure 1.78 *Circuit for determining resistance*

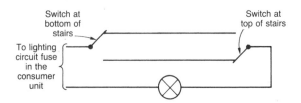

Figure 1.79 *Two-way lighting switch*

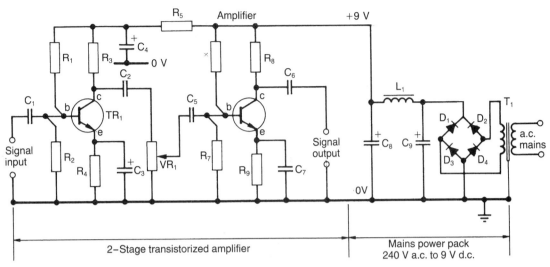

Figure 1.80 *A two-stage transistor amplifier*

Table 1.3 *Component list for the two-stage transistor amplifier*

Component	Description
R_1–R_9	Fixed resistors
VR_1	Variable resistor
C_1–C_9	Capacitors
D_1–D_4	Diodes
TR_1, TR_2	Transistors
T_1	Mains transformer
L_1	Inductor (choke)

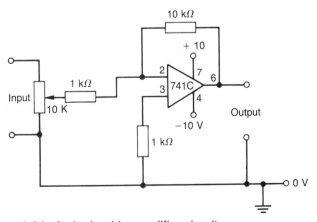

Figure 1.81 *A single-chip amplifier circuit*

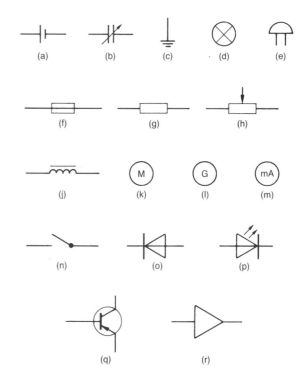

Figure 1.82 *See Test your knowledge 1.25*

Test your knowledge 1.25

Figure 1.82 shows some typical electrical and electronic component symbols. Name them and briefly explain what they do.

Activity 1.10

Draw a simple circuit for powering four electric lamps from a battery. It must be possible to turn each lamp on or off independently. A master switch must also be provided so that the whole circuit can be turned on or off. A fuse must be included in the circuit in order to prevent the battery being overloaded.

Activity 1.11

Draw a schematic circuit diagram for a battery charger having the following features:
- the primary circuit of the transformer (i.e. the side that is connected to the a.c. mains supply) is to have an on/off switch, a fuse and an indicator lamp
- the secondary circuit of the transformer is to have a bridge rectifier, a variable resistor to control the charging current, a fuse and an ammeter to indicate the charging current.

Figure 1.83 shows an electronic circuit.
(a) Draw up a component list that numbers and names each of the components (include values where given).
(b) Suggest what the circuit might be used for (Hint: the circuit only has an output!)

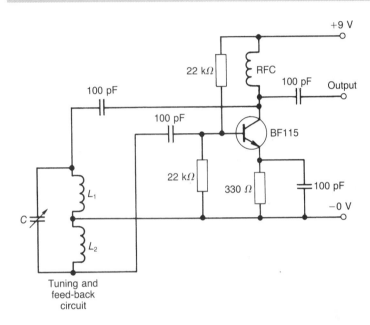

Figure 1.83 *See Activity 1.12*

Design

An engineered product or service is, by definition, one that is designed. The design of an engineered product has to take into account a large number of factors including the shape, dimensions, tolerances, materials, and finish of the product. We shall begin by looking at the initial brief that is required before any design work can begin before moving on to look at the other stages in the design process.

The design brief

A design brief takes the form of an instruction to a designer (or a design team) specifying a requirement for the design. A design brief may be the outcome of discussions with a customer or client or it may arise out of market research which has identified the need for a new product. In either case, design briefs can vary considerably in both form and content.

A design brief can also arise out of the need to improve the design of an existing product. Many companies find it more cost effective to refine existing products and services rather than to design new products and services. By modifying and improving on an existing design it can be possible to reduce manufacturing costs, improve reliability, enhance maintainability, and meet any new legal standards that may come into force.

The key features of a design brief are:

- Application (what the product is used for *and* who will be using it).
- Technical specification (the required performance for the product).
- Ergonomic considerations (including ease of use and adaptability to suit different users).
- Aesthetic considerations (details on styling, general appeal, range of colour options, etc.).
- Quality (and how it is related to the cost of the product).
- Aesthetics (appearance and styling of the product).
- Size (specified in all three dimensions).
- Weight (size and weight are often *very* important to customers and end users of the product).
- Maintenance (how this is planned for in the design and subsequently during the life of the product).
- Materials (and the manufacturing processes required).
- Cost (including design, production and material costs).
- Legislation (regulations including those relating to health and safety).
- Scale of production (quantity required and the timescale for delivery).

Constraints

The constraints that may impinge on the design brief include:

- technological
- resources (labour, materials, plant)
- environment
- cost.

In most cases it is necessary to work well within the bounds of existing technology while recognizing that some design briefs may demand technological development before implementation.

The brief need not necessarily apply to a new invention, it is quite acceptable to redesign part of a product, for example, to reduce the number of components or take advantage of new materials and technology. Having checked the customer requirements, standards and legislation and the constraints imposed on you, you should now be able to produce a design brief.

A design brief is actually a description of the customers needs, so it is vital to get the design brief correct. When working on your own design as part of your GNVQ course you will need to discuss the brief with your 'customer'. Ideally this will be someone in industry but it may be your lecturer or tutor or it could be someone that you envisage as being the 'end user' of your product. In any event, you should use the listed considerations as a 'ticking list' to check that you have covered all of the aspects of the design brief. Make absolutely certain that the brief *clearly and unambiguously* states the customer requirements!

Standards and legislation

Relevant standards and current legislation affect every engineered product. The considerations that may apply include:

- Health and Safety legislation
- codes of practice
- conventions
- British Standards
- European Union directives
- International Standards
- internal company standards.

Standards and legislation varies widely according to the type of product or service. You are expected to be aware of any legislation relevant to the product or service you are studying, you are advised to consult the latest BSI *(British Standard Institute)* catalogue for the relevant standards and information you need.

Feasibility studies

The next stage is a feasibility study of the design brief. This entails an in-depth study of the brief to establish whether a feasible design is possible to satisfy all the requirements the customer expects, it is at this stage all sorts of problems need consideration. Sometimes these problems will require a compromised solution agreed between you and the customer.

By the time you have completed the feasibility study and found proposed solutions to satisfy the design brief, many ideas and solutions are experienced, remember though at this stage the task has only been to see whether the design brief is feasible, and probably very little real design work has taken place, apart from investigations and design possibilities.

The problems most generally encountered in engineered products are perhaps those associated with weight or size, and other problems may occur by not being able to keep pace with changes in technology, especially electronic devices, and you could find your design is outdated 'before it has left the ground'.

When you are confident that you will be able to further the ideas expressed in the design brief from the customer and you are satisfied your feasibility study shows you would be capable of handling the design, the next stage is to submit to your customer a design proposal.

Design proposals

By proposing solutions, a greater understanding of the overall design problem will be gained which in itself will show many sub-problems. The design proposal can take a considerable amount of time and money to investigate and produce, it will need to convince the customer that you are, as a company, capable of solving the problems of design, and have the expertise and qualified staff to do so.

On an elaborate equipment that requires knowledge that is close to the edge of technology, for instance, a ship's complete radar system, the design proposal can cost thousands of pounds and take months to complete. If the contract demands a competitive tender, as it almost certainly would, then even more time will be spent to estimate the cost of the contract, which is a difficult task as the final design specification has yet to be agreed.

The design proposal sent to the customer basically consists of:

- a provisional design specification
- proposed overall contract costs
- delivery dates
- servicing and maintenance
- any legal requirements.

Design solutions

Having dealt at some length with design briefs and design specifications, we can now consider how you should go about handling a design specification (assuming you are now prepared to go ahead with your proposed design). At this stage, it is worth remembering that the design specification is *not* a design solution.

Having established that you now have a reasonable insight into the design requirements of the project specification, we can set up a design programme of the project from start to finish. However, before designing begins we need to consider what design strategy should be adopted.

There are basically six stages in the design process:

1 Clarifying objectives.
2 Establishing functions.
3 Setting requirements.
4 Generating alternatives.
5 Evaluating alternatives.
6 Improving details.

Let's look in a little more detail at the stages that make up the design process and expand the meaning of each:

1 Clarifying objectives
  This stage would in all probability have been dealt with during the design brief and design proposals stages, when the objectives of the design would be stated (but you will still need to clarify them before you actually begin designing!).

  Aim
  To clarify the design objectives and sub-objectives, and the relationships between them.

  Method
  This is best achieved by discussions with the design team and

questions to the customer.

2 Establishing functions

Although the customer may have specified the functions expected from the design the designer may find a more radical or innovative solution by reconsidering the level of the problem definition. He or she may be able to offer the customer a better solution to the functional problems of the design in excess to the expected at no extra cost.

Aim

To establish the functions required.

Method

Break down the overall function into sub-functions. The sub-functions will comprise of all the functions expected within the product.

3 Setting requirements

Design problems are usually all set within certain limits, these limits may be cost, weight, size, safety or performance, etc. or any combination of them.

Aim

To produce an accurate specification of the performance of the designed product.

Method

Identify the required performance attributes, these may well have been considered at the design feasibility study stage.

4 Generate alternatives

Even if you think you have a good design solution always look further, if time and costs permit for an alternative solution, hopefully better than the one you thought would be an ideal design.

Aim

To have a choice of solutions to allow comparisons of ideas in solving the design requirements.

Method

It would help to draw several design layout drawings to enable discussions to take place with the design team and the customer.

5 Evaluate alternatives

When some alternative design proposals have been thought about and maybe some design layouts have been produced, the evaluation of the alternatives can be discussed.

Aim

To evaluate the alternatives, choose the ones that satisfy the customer requirements and are good value to him.

Method

Compare the value of the alternative design proposals against the original proposal agreed with the customer on the basis of performance.

Test your knowledge 1.26

Prepare an outline design brief for a multi-purpose tool suitable for use by campers and hikers. Suggest uses for the tool and prepare a number of sketches showing alternative design solutions.

Test your knowledge 1.27

Prepare an outline design brief for a portable kitchen timer. Suggest uses for the product and prepare a number of sketches showing alternative design solutions.

6 Improving details
There are mainly two reasons for improving details, they are either aimed at increasing the product value to the customer or reducing the cost to the producer.

Aim
To increase or maintain the value of the product to the customer at the same time reducing its cost to the producer.

Method
This can be approached using two methods, one is called value engineering and the other value analysis.

Communicating your designs

Almost certainly you will need to communicate your design by drawing it, as this is by far the easiest way to describe your intention, in fact to verbally describe a design can cause confusion as the persons listening may well imagine something totally different to the idea as you perceive it.

A knowledge of a variety of methods used for graphical communication is essential if you are to successfully submit your design ideas. In particular you should be able to:

- select appropriate graphical methods to be used for communicating engineering information
- produce scale and schematic drawings for engineering applications
- interpret information presented in engineering drawings.

Generally you will need sufficient drawings to fully explain your design concept, this means an ability to draw the following, if required:

- Layout drawings (the original sketches and drawings required to show your design proposals)
- Detail drawings (dimensioned drawings of any manufactured parts)
- Assembly drawings (showing how the product should be assembled)
- Item lists (listing all component parts required to make the final assembly)

Although drawings should normally be produced to British Standard BS 308, you may find sketches drawn in good proportion are acceptable, you will need to ask your tutor. Also note that initial design concepts are often made up from rough sketches.

It is important to ensure that your drawings or sketches can be understood easily and without confusion. In particular, you should remember to state the materials and finishes required on any of the parts to be drawn and be careful with your dimensioning and tolerancing. In a manufacturing environment, it is important to obtain approval for each drawing from a qualified production engineer.

He or she will be able to advise on all aspects of the design that might have an impact on the production process, including the suitability of the various production processes.

A designer's checklist

The following checklist will help you to develop your own designs:

Design brief

Is the design brief adequate? Have you understood what is required? Are there any questions that you need answered? Have all the constraints been identified?

Design specification

Has the specification been agreed? Is it comprehensive? Does the specification cover all features mentioned in the design brief? Are there any legal issues or standards that must be complied with? How can you ensure that the specification is met?

Material suitability

Have the most appropriate materials been selected? What processes can be applied most cost-effectively to this material? Is the material machinable and/or weldable if required? Could machining time be saved by using stock sizes? (check raw material tolerances)

Dimensioning

Check that there are sufficient dimensions to manufacture the item and the drawing can be clearly understood. Dimensioning drawings from left to right is good drawing office practice, check that no dimensions are left to machine shop calculation. Ensure inside and/or outside radii are stated.

Datums

Datums are preferred if made from vertical and horizontal edges or a datum hole, the fewer datums the better (holes can be used for tooling purposes).

Tolerances

Note the tolerances specified, if in doubt enquire whether any tight tolerances stated are justified. Wider tolerances could save considerably on manufacturing costs (often there can be a misuse of geometric tolerancing).

Machined finishes

Surface texture is often specified on a drawing, check the necessity of any very fine finishes shown, by establishing the function of the component part.

Heat treatments

Check the heat treatment specification (if specified) is correct for the material.

Coated finishes

Check the specified coated finish is applicable to the material used.

Machine processes

Could the shape be slightly changed to allow for easier machining, if the answer is yes, consultation with the designers will be necessary.

Assemblies and sub-assemblies

Check whether the number of parts could be reduced by using standard parts, re-design or machining from the solid. Check the build up of tolerances on assemblies.

Other points

Do the final assemblies satisfy any interchangeability requirements? Is the item commensurate with the layout or scheme previously agreed in conjunction with the designers?

Notes:

At the initial stage, agreement must be achieved with regard to the technique and method of manufacture, materials, critical dimensions, environment, conditions of use, etc. These should *all* have been satisfied and agreed in advance.

Finally, it is important to evaluate your final design solution against the customer's original specification — only by doing this will you be able to improve your skills as a designer!

Activity 1.13

Produce a full design specification for a nickel cadmium (NiCd) battery charger. The charger is to accept various sizes of NiCd battery and apply an appropriate charging current to the cell. The charger should incorporate a visual indicator to show that a battery is connected and is charging. As part of this Activity you will need to obtain literature on NiCd cells and investigate any other similar products that are currently available to meet this need. You should present your design work as a set of initial design solutions, including sketches and drawings.

Activity 1.14

Select and present a final design solution for the NiCd battery charger. The following items will be required: detailed assembly drawings, a complete circuit diagram, a PCB layout, an internal wiring diagram, and a detailed parts list. Present these in the form of a design portfolio suitable for passing to a production engineer prior to manufacture.

Review questions

1   List THREE main considerations that should be taken into account when producing a design brief.

2   Explain the purpose of a design proposal.

3   Produce an outline design specification for a portable electric screwdriver.

4   Describe THREE different types of diagram that may be used as part of an engineering design solution. Illustrate your answer with sketches.

5   Identify THREE constraints that are likely to have an impact on a design brief. Illustrate your answer by giving a typical example.

6   Explain what is meant by a feasibility study in relation to a design brief.

7   Describe the sequence of stages in arriving at a design for an electric screwdriver suitable for home use. Use a flowchart to illustrate your answer.

8   Sketch typical line styles used to illustrate:
(a) a centre line
(b) the limit of a partial or interrupted view.

9   How many A2 drawing sheets can be cut from an A0 drawing sheet? Explain your answer with a sketch.

10   State FOUR items that should be included within the title block of an engineering drawing.

11   Sketch engineering drawing symbols that are used to indicate:
(a) a 4/2 directional control valve
(b) a non-return valve
(c) a battery
(c) a variable resistor
(d) a semiconductor diode
(e) an iron-cored transformer.

12   Draw, using appropriate symbols, a two-way lighting circuit. Label your drawing clearly.

13   List FOUR advantages of using CAD in the preparation of engineering drawings compared with purely manual methods.

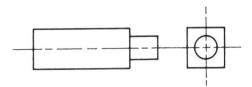

Figure 1.84

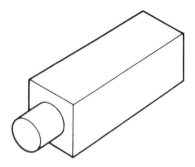

Figure 1.85

14   Identify the projections used in Figures 1.84 and 1.85.

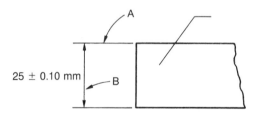

Figure 1.86

15   Identify the lines marked A and B in Figure 1.86.

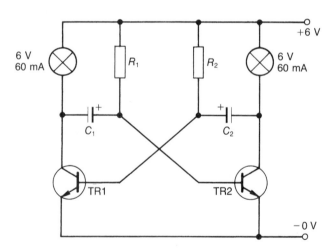

Figure 1.87

16   Identify each of the components shown in Figure 1.87.

Unit 2 Application of new technology in engineering

Summary

The use of new technology is an exciting and expanding part of engineering. New technology is used in all sectors of engineering and will continue to develop in the foreseeable future. In this unit you will look at the role of new technology through a series of short case studies. Each case study focuses on a different product or service and shows how new technology has made the development possible. Each case study introduces one or more aspects of the application of the three main areas of new technology; information technology, new materials and components, and automation.

The case studies introduced in this chapter contain a number of activities for you to complete. These are designed to provide you with real-life examples of the development and application of new technology. They will also provide you with opportunities to develop your skills in gathering and using information. In order to complete the case study assignments you will need to make use of your school or college library as well as other information sources such as CD-ROMs and the World Wide Web.

This unit has links with several other intermediate units, particularly those that involve the use of computer-aided design and making engineered products. You should look out for ways of using your knowledge of new technology in these units also!

Information technology

Information technology is not just about computers — it's about all aspects of accessing, processing and disseminating information. This revolution has come about as a result of the convergence of several different technologies, notably:

- computing (microprocessors, memories, magnetic and optical data storage devices, etc.)
- telecommunications (data communications, networking, optical fibres, etc.)

- software (optical character recognition, data transfer protocol, visually orientated programming languages, etc.).

The first two case studies in this section deal with data sources and data handling whilst the final case study is about computer aided engineering (CAE).

Case study

Databases

A database is simply an organized collection of data. This data is usually organized into a number of records each of which contains a number of fields. Because of their size and complexity and the need to be able to quickly and easily search for information, a database is usually stored within a computer and a special program – a *database manager* or *database management system* (DBMS) – provides an interface between users and the data itself. The DBMS keeps track of where the information is stored and provides an index so that users can quickly and easily locate the information they require (see Figure 2.1).

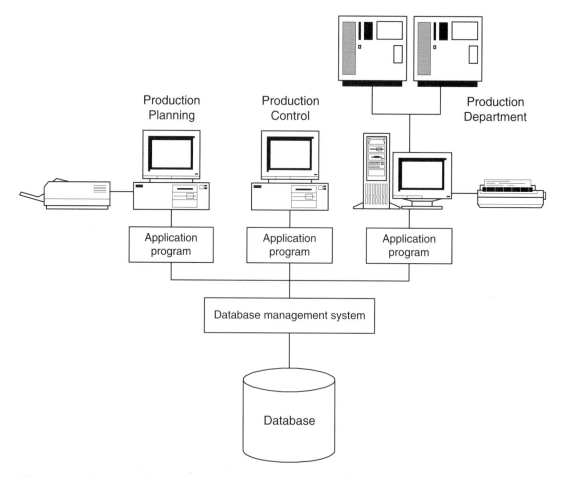

Figure 2.1 *A database management system (DBMS)*

The database manager will also allow users to search for related items. For example, a particular component may be used in a number of different products. The database will allow you to quickly identify each product that uses the component as well as the materials and processes that are used to produce it.

The structure of a simple database is shown in Figure 2.2. The database consists of a number of *records* arranged in the form of a table. Each record is divided into a number of *fields*. The fields contain different information but they all relate to a particular component. The fields are organized as follows:

> Field 1 Key (or index number)
> Field 2 Part number
> Field 3 Type of part
> Field 4 Description or finish of the part

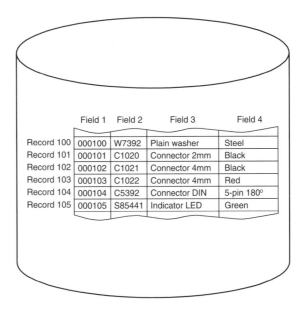

Figure 2.2 *The structure of a simple parts database*

The database management system

The database management system is used to build and maintain the database. It also provides the interface between the user and the information stored in the database. Tasks performed by the DBMS include:

- adding new records
- deleting unwanted records
- amending records
- linking or cross-referencing records
- searching and sorting the database records
- printing reports of selected records.

Types of database

We have already said that many companies use specialized databases in order to manage different functions within the company. To help you understand this, consider the case of an engineering company that manufactures forklift trucks. The company might use the following databases to help it organize the different aspects of its operation:

- a Product Database containing records of each vehicle manufactured and any modifications fitted
- a Manufacturing Database containing records of all components and materials used during manufacture
- a Customer Database containing records of all customers and the vehicles supplied to them
- a Spare Parts Database containing records of all spare parts held showing where they are stored and the quantity held.

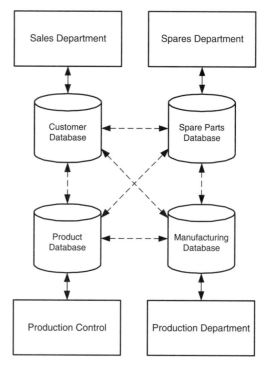

Figure 2.3 *Four specialized databases and their relationships with different company functions*

Figure 2.3 shows how these four databases relate to several different departments within the company. Since these company functions cannot operate in isolation there is a need to exchange data between the three databases. To give you some idea of how this might work, let's assume that the Sales Department has been given the task of marketing a new forklift truck that has just become available. The Manager of the Sales Department has been given a target that requires him to sell 10 forklift trucks by the end of the next quarterly period. He has been allocated a fixed budget specifically to meet the costs associated with this sales campaign.

The Sales Manager decides to produce a brochure giving details of the new forklift truck and offering a substantial discount to any previous customers who may wish to 'trade-in' their existing fleet of forklift trucks. The brochure will be mailed to all UK customers that have purchased forklift trucks in the last ten years. Sales staff will then follow this up with a telephone call to each named customer contact.

The information required to produce the new forklift truck brochure will be drawn from the company's Product Database. The mailing list, contact names and telephone numbers will be taken from a report generated from the Customer Database.

The structure and content of a typical record in the Product Database will include the following information:

> Product reference number: FLT1022
> Product name: Challenger
> Vehicle chassis type: BCX077
> Quantity in stock: 7
> Scheduled production (current period):   10
> Scheduled production (next period):   12

The structure and content of a typical record in the Customer Database will include the following information:

> Customer reference number: 13871
> Company name: Enterprise Air Freight
> Address (line 1):   Unit 8
> Address (line 2):   Bath Road Industrial Estate
> Town/city: Feltham
> County or state: Middlesex
> Post code or zip code: UB10 3BY
> Country: UK
> Contact name: David Evans
> Contact title: Purchasing Manager
> Contact salutation: Dear David
> Telephone number: 0181-979-7756
> Fax number: 0181-979-7757
> e-mail: david.evans@enterprise.co.uk

Test your knowledge 2.1

In setting up the mailshot that will be used to sell the fork lift trucks (see text) what fields within the Customer Database will be:

(a) used to determine which customers are included in the mailing list

(b) used to generate the address label that will be attached to the information pack.

Activity 2.1

Investigate the use of databases in your school or college. Working as part of a group, begin by finding out the name of the person who has overall responsibility for collecting and processing student data. Interview him or her and find out what information is held in the database and how it is organized. Also find out about the reports that are generated by the database and who has access to the information. Prepare a brief presentation to the rest of your class using appropriate handouts and visual aids.

There is a need to ensure that information contained in a database is kept up-to-date. In the case of the customer database, which records and field may need updating as a result of the mailshot and telephone sales campaign?

Nowadays, there is a trend towards integrating many of the databases within an engineering company into one large database. This database becomes central to all of the functions within the company. In effect, it becomes the 'glue' that holds all of the departments together.

The concept of a centralized manufacturing database is a very sound one because it ensures that every function within the company has access to the same data. By using a single database, all departments become aware of changes and modifications at the same time and there is less danger of data becoming out-of-date.

Case study   The Internet and the World Wide Web

Although the terms *Web* and *Internet* are often used synonymously, they are actually two different things. The *Internet* is the global association of computers that carries data and makes the exchange of information possible. The *World Wide Web* is a subset of the Internet – a collection of inter-linked documents that work together using a specific Internet protocol called *hypertext transfer protocol* (HTTP). In other words, the Internet exists independently of the World Wide Web, but the World Wide Web can't exist without the Net.

The World Wide Web began in March 1989, when Tim Berners-Lee of the European Particle Physics Laboratory at CERN (the European centre for nuclear research) proposed the project as a means to better communicate research ideas among members of a widespread organization.

Web sites are made up of collections of Web pages. Web pages are written in *hypertext markup language* (HTML), which tells a *Web browser* (such as Netscape's Communicator or Microsoft's Internet Explorer) how to display the various elements of a Web page. Just by clicking on a *hyperlink*, you can be transported to a site on the other side of the world.

A set of unique addresses is used to distinguish the individual sites on the World Wide Web. An Internet Protocol (IP) address is a 4– to 12–digit number that identifies a specific computer connected to the Internet. The digits are organized in four groups of numbers (which can range from 0 to 255) separated by full stops. Depending on how an Internet Service Provider (ISP) assigns IP addresses, you may have one address all the time or a different address each time you connect.

Every Web page on the Internet, and even the objects that you see displayed on Web pages, has its own unique address, known as a *uniform resource locator* (URL). The URL tells a browser exactly where to go to find the page or object that it has to display.

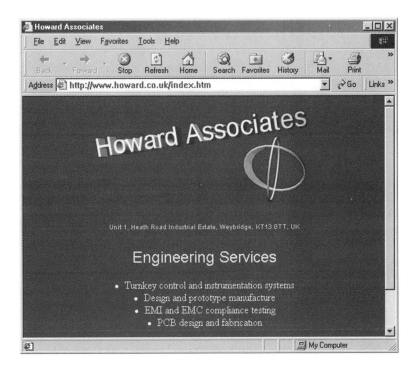

Figure 2.4 *A typical engineering company's Web site displayed in a Web browser*

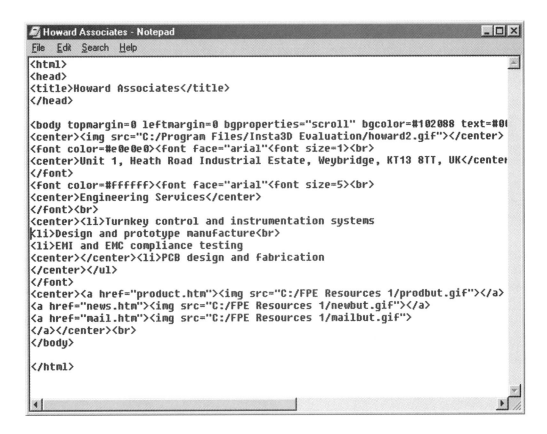

Figure 2.5 *The HTML code responsible for generating the page shown in Figure 2.4*

On-line services

On-line services can be defined as services that add value to the World Wide Web. Originally, these services built and maintained trunk networks that could be used by their customers. They also added their own *content* (such as news and weather reports, software libraries, etc.) to the Web. Users who are prepared to pay for the service can access this material.

Customers still pay for some on the on-line services but the trend, in recent years, has been to make added value services free. Currently the most popular on-line services are America Online (AOL), CompuServe, Prodigy, and the Microsoft Network (MSN). All of these services provide access to e-mail, support libraries, and on-line communities where people with similar interests can communicate for business or pleasure.

Search sites

Being able to locate the information that you need from a vast number of sites scattered across the globe can be a daunting prospect. However, since this is a fairly common requirement, a special type of site, known as a *search site,* is available to help you with this task. There are three different types of search site on the Web; *search engines*, *Web directories*, and parallel and *metasearch sites*.

Search engines such as Excite and HotBot use automated software called *Web crawlers* or *spiders.* These programs move from Web site to Web site, logging each site title, URL, and at least some of its text content. The object is to hit millions of Web sites and to stay as current with them as possible. The result is a long list of Web sites placed in a database that users search by typing in a keyword or phrase.

Web directories such as Yahoo and Magellan offer an editorially selected, topically organized list of Web sites. To accomplish that goal, these sites employ editors to find new Web sites and work with programmers to categorize them and build their links into the site's index.

To make things even easier, all the major search engine sites now have built-in topical search indexes, and most Web directories have added a keyword search.

Intranets

Intranets work like the Web (with browsers, Web servers, and Web sites) but companies and other organizations use them internally. Companies use them because they let employees share corporate data, but they're cheaper and easier to manage than most private networks—no one needs any software more complicated or more expensive than a Web browser, for instance.

They also have the added benefit of giving employees access to the Web. Intranets are closed off from the rest of the Net by firewall software, which lets employees surf the Web but keeps all the data on internal Web servers hidden from those outside the company.

Extranets

One of the most recent developments has been that of the *extranet.* Extranets are several intranets linked together so that businesses can

---

Test your knowledge 2.3

What do each of the following abbreviations stand for?

(a) IP
(b) HTML
(c) URL
(d) ISP.

Test your knowledge 2.4

Explain why sites on the World Wide Web must all have unique addresses.

Test your knowledge 2.5

Name two Web browsers and explain how they are used to locate and view a Web site.

share information with their customers and suppliers. Consider, for example, the production of a European aircraft by four major aerospace companies located in different European countries. They might connect their individual company intranets (or parts of their intranets) to each other, using private leased lines or even using the public Internet. The companies may also decide to set up a set of private newsgroups so that employees from different companies can exchange ideas and share information.

Activity 2.2

Visit the Web site of Lansing Linde UK, a manufacturer of forklift trucks. Investigate the company and view some of the gallery of photographs that show the development of the company's forklift trucks. Search the site for information on its latest range of electric forklift trucks capable of handling loads from 1,000 kg to 8,000 kg. Write a brief report describing these trucks and make specific reference to features that:

- save energy
- ensure smoother driving
- protect the driver
- improve visibility
- facilitate turning and manoeuvring
- ensure stability.

The URL of the Lansing Linde Web site is:

http://www.lll.co.uk

Activity 2.3

Use a search engine (such as Lycos or AltaVista) to locate information about DVD-ROMs. Visit the first four sites displayed as a result of your search and note down the URL of each of these sites. Summarize the contents of each of the sites by writing a paragraph describing each site. Then rate each site on a scale of 1–10 on the basis of content, presentation, and ease of use. Summarize your results in a table. Repeat the activity using a Web directory (such as Yahoo or Excite). Use the directory to navigate to four different sites giving details of DVD-ROM. Once again, note down the URL of each site, summarize its contents and rate it on a scale of 1–10 (again presenting your results in the form of a table). Compare these two search methods.

Activity 2.4

Set up a Web-based e-mail account in your own name. Note down all of the steps that you took to open the account including details of any electronic forms that you had to complete. Present your findings in the form of a brief article for your local paper on how to open and use an e-mail account. You should assume that the reader is non-technical.

Case study

Computer aided engineering

You will probably already know a little about computer-aided design (CAD) however this is just one aspect of computer aided engineering (CAE). Computer aided engineering is about automating *all* of the stages that go into providing an engineered product or service.

When applied effectively, CAE ties all of the functions within an engineering company together. Within a true CAE environment, information (i.e. data) is passed from one computer aided process to another. This may involve computer simulation, computer aided drawing (drafting), and computer aided manufacture (CAM).

The single term, CADCAM, was once used to describe the integration of computer aided design, drafting and manufacture. Another term, CIM (computer integration manufacturing), is often applied to an environment in which computers are used as a common link that binds together the various different stages of manufacturing a product, from initial design and drawing to final product testing.

Whilst all of these abbreviations can be confusing (particularly as some of them are often used interchangeably) it is worth remembering that 'computer' appears in all of them. What we are really talking about is the application of computers within engineering. Nowadays, the boundaries between the strict disciplines of CAD and CAM are becoming increasingly blurred.

We have already said that CAD is often used to produce engineering drawings. Several different types of drawing are used in engineering. Some examples are shown in Figures 2.6–2.9.

Computer aided manufacture (CAM) covers a number of more specialized applications of computers in engineering including computer integrated manufacturing (CIM), manufacturing system modeling and simulation, systems integration, artificial intelligence applications in manufacturing control, CAD/CAM, robotics, and metrology.

Computer aided engineering analysis can be conducted to investigate and predict mechanical, thermal and fatigue stress, fluid flow and heat transfer, and vibration/noise characteristics of design concepts to optimize final product performance. In addition, metal and plastic flow, solid modeling, and variation simulation analysis are performed to examine the feasibility of manufacturing a particular part.

In addition, all of the machine tools within a particular

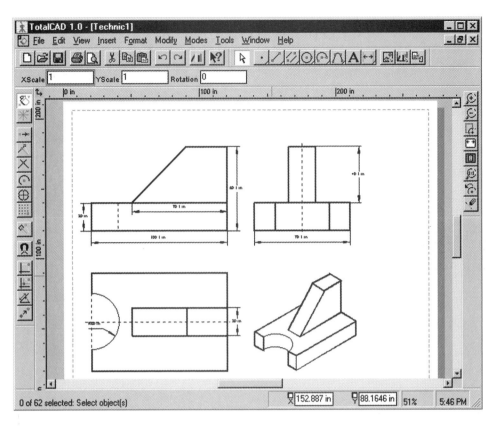

Figure 2.6 *A conventional engineering drawing produced by a CAD package*

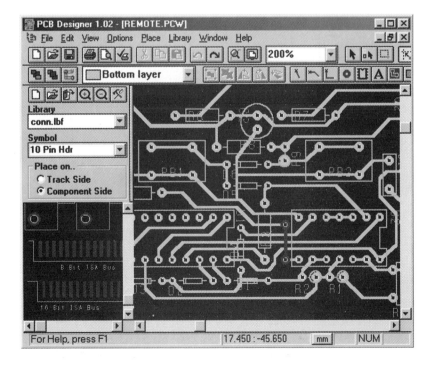

Figure 2.7 *Printed circuit board design is another excellent application for CAD*

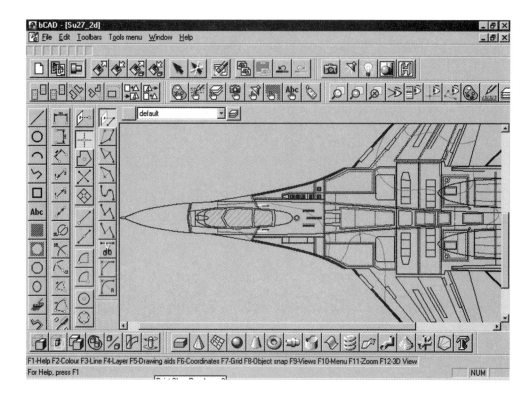

Figure 2.8 *Complex drawings can be produced prior to generating solid 3-D views*

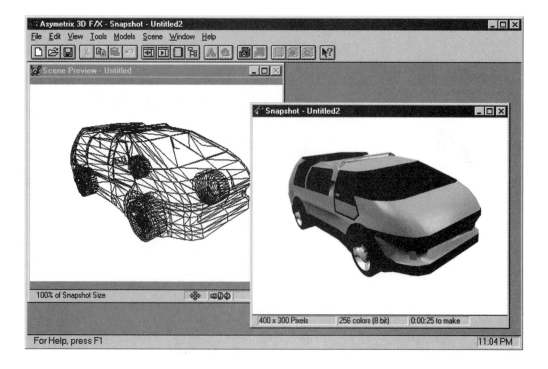

Figure 2.9 *A wire frame 3-D drawing and its corresponding rendered view*

manufacturing company may be directly to the CAE network through the use of centrally located floor managers which monitor machining operations and provides sufficient memory for complete machining runs.

Manufacturing industries rely heavily on computer controlled manufacturing systems. Some of the most advanced automated systems are employed by those industries that process petrol, gas, iron and steel. The manufacture of cars and trucks frequently involves computer-controlled robot devices. Industrial robots are used in a huge range of applications that involve assembly or manipulation of components.

Another development that has greatly affected the manufacturing industries is the integration of engineering design and manufacturing into one continuous automated activity through the use of computers. The introduction of CAD/CAM (computer aided design and computer aided manufacturing) has significantly increased productivity and reduced the time required to develop new products. When using a CAD/CAM system, an engineer develops the design of a component directly on the display screen of a computer. Information about the component and how it is to be manufactured is then passed from computer to computer within the CAD/CAM system.

After the design has been tested and approved, the CAD/CAM system prepares instructions for computer numerically controlled (CNC) machine tools and places orders for the required materials and an additional parts (such as nuts, bolts or adhesives). The CAD/CAM system allows an engineer (or, more likely, a team of engineers) to perform all the activities of engineering design by interacting with a computer system (invariably networked) before actually manufacturing the component in question using one or more CNC machines linked to the CAE (computer aided engineering) system.

Test your knowledge 2.6

What do each of the following abbreviations stand for?

(a) CAD
(b) CAM
(c) CADCAM
(d) CIM
(e) CNC
(f) CAE.

Activity 2.5

Use a CAD package, such as Autosketch, to produce a 2-D plan of the workshop area in your school or college. Choose an appropriate scale and ensure that your plan is fully dimensioned. Mark on your plan the location of machine tools and other equipment available. Print or plot the finished plan and ensure that you have adopted the recommended style, framing and presentation conventions used in your school or college.

Telecommunications

Telecommunications provides the infrastructure that allows information to be transferred from place to place. The term includes communications by conventional wires (or 'lines'), cables, radio microwave and optical technology.

Case study	# Optical fibres

An optical fibre is a long thin strand of very pure glass enclosed in an outer protective jacket. Because the refractive index of the inner layer is larger than that of the outer layer, light travels along the fibre by means of total internal reflection at a speed of around 200 million m/s (approximately 2/3 of the speed of light).

In order to set up an optical data link, the optical fibre must be terminated at each end by means of a transmitter/receiver unit. A simple one-way fibre optic link is shown in Fig. 2.10.

The optical transmitter consists of a light emitting diode (LED) coupled directly to the optical fibre. The LED is supplied with pulses of current from a computer interface. The pulses of current produce pulses of light that travel along the fibre until they reach the optical receiver unit.

The optical receiver unit consists of a photodiode (or phototransistor) that passes a relatively large current when illuminated and hardly any current when not. The pulses of current at the transmitting end are thus replicated at the receiving end.

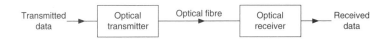

Figure 2.10 *A one-way (unidirectional) optical fibre data link*

A bi-directional optical fibre link can be produced by adding a second fibre and having a transmitter and receiver at each end of the link (Fig. 2.10). A simple optical fibre data link of this sort is capable of operating at data rates of up to 500 kb/sec and distance of up to 1 km. More sophisticated equipment (using high-quality low-loss fibres) can work at data rates of up to 250 Mb/s and at distances of more than 10 km.

What type of semiconductor device is used for:

(a) an optical receiver
(b) an optical transmitter.

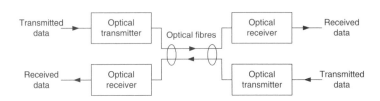

Figure 2.11 *A two-way (bidirectional) optical fibre data link*

Activity 2.6

There are several different types of optical fibre. Identify at least three different types and explain why they are different. Present your findings in the form of a series of word-processed data sheets.

Activity 2.7

Optical fibres can provide a bandwidth that is much greater than that which can be obtained using conventional copper cables. Write a brief word-processed report explaining why this is.

Case study

GPS receivers

The Global Positioning System (or GPS) is a collection of satellites owned by the U.S. Government that provides highly accurate, worldwide positioning and navigation information, 24 hours a day. It is made up of 24 NAVSTAR GPS satellites that orbit 12,000 miles above the earth, constantly transmitting the precise time and their position in space. GPS receivers on (or near) the surface of the earth, listen in on the information received from three to twelve satellites and, from that, determine the precise location of the receiver, as well as how fast and in what direction it is moving.

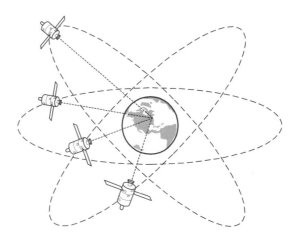

Figure 2.12 *GPS system using triangulation—a minimum of three satellites are required to determine position*

GPS uses the triangulation of signals from the satellites to determine locations on earth (see Fig. 2.12). GPS satellites know their location in space and receivers can determine their distance from a satellite by using the travel time of a radio message from the satellite to the receiver. After calculating its relative position to at least three or four satellites, a GPS receiver can calculate its position using triangulation. GPS satellites have four highly accurate atomic clocks on board. They also have a database (sometimes referred to as an 'almanac') of the current and expected positions for all of the satellites that is frequently updated from earth. That way when a GPS receiver locates one satellite, it can download all satellite location information, and find the remaining needed satellites much more quickly.

Over the last several years, an increasing variety of affordable GPS receivers have been released for the average consumer. As the technology has improved, many additional features are added to these units, while the price and size continue to decrease.

Some of the more specialized GPS receivers currently available include:

- handheld GPS receivers that have background maps
- GPS receivers fitted to cars and lorries (with integral databases of maps and road information)
- GPS receivers for large and small boats (including those integrated with other navigational equipment)
- aircraft GPS receivers with built-in airport information
- GPS receivers that combine with Internet access and/or e-mail into one unit.

GPS products have been developed for use for many commercial applications including surveying, map making, tracking systems, navigation, construction and mapping natural resources. The recreational use of GPS includes sea fishing, hiking, skiing and mountain walking.

Activity 2.8

Obtain the data sheets and/or technical specification for at least two different types of low-cost GPS receiver. Identify the market for these receivers and write a brief article for your local newspaper explaining what GPS can do for ordinary readers.

Test your knowledge 2.8

What is the minimum number of satellites within sight of a GPS receiver are needed to fix a position? Explain your answer.

Activity 2.9

Visit a local car showroom and investigate the GPS systems that may be supplied with the latest models. Write a brief article for a car enthusiasts' magazine explaining, in simple terms, how GPS works and how it can benefit the car driver.

New materials and components

Ongoing developments in materials and materials processing has made possible a number of exciting new products. The case study in this section is about a group of materials commonly referred to as plastics but more correctly called polymers. Because of their unique properties, these materials are now widely used in engineering.

Case study

Polymers

Polymers are large molecules made of small, repeating molecular building blocks called *monomers*. The term polymer is a composite of the Greek words *poly* and *meros,* meaning 'many parts'. The process by which monomers link together to form a molecule of a relatively high molecular mass is known as *polymerization.*

Polymers are found in many natural materials and in living organisms. For example, proteins are polymers of amino acids, cellulose is a polymer of sugar molecules, and nucleic acids such as deoxyribonucleic acid (DNA) are polymers of nucleotides. Many man-made materials, including nylon, paper, plastics, and rubbers, are also polymers. Plastics are often divided into two main groups; *thermoplastics* and *thermosetting plastics* (or *thermosets*).

There is an important difference between these two classes of material. Because themosetting plastics undergo a chemical change during moulding, they cannot be softened by reheating. Thermoplastics, on the other hand, become soft whenever they are reheated. Thermoplastics are softer and more pliable than thermosetting plastics which tend to be hard and brittle.

The chemical change experienced by thermosetting plastic materials (referred to as called *curing*) is brought about from the temperature and pressure that is applied during the moulding process.

The properties of some common thermoplastics are shown in Table 2.1 whilst the properties of some common thermosetting polymers are given in Table 2.2.

Test your knowledge 2.9

Explain the difference between thermoplastics and thermosets.

Test your knowledge 2.10

Which polymer material listed in the table is most suitable for:

(a) providing a non-stick coating in a cooking pan
(b) insulating the mains lead of a soldering iron
(c) manufacturing the outer moulded case of a portable compact disk player.

Give reasons for your answers.

Table 2.1 *Properties of some common thermoplastics*

Material	Relative density N/mm^2	Tensile strength J	Elongation %	Impact strength	Max. service temperature °C
Polyamide (nylon)	1.14	50–85	60–300	1.5–15	120
Polythene	0.9	30–35	50–600	1–10	150
Polypropylene	1.07	28–53	1–35	0.25–2.5	65–85
Polystyrene	1.4	49	10–130	1.5–18	70
PVC	1.3	7–25	240–380	–	60–105
Teflon (PTFE)	2.17	17–25	200–600	3–5	260

Table 2.2 *Properties of some common thermosetting polymers*

Material	Relative density N/mm^2	Tensile strength J	Elongation %	Impact strength	Max. service temperature °C
Epoxide	1.15	35–80	5–10	0.5–15	200
Urea formaldehyde	1.50	5–75	1.0	0.3–0.5	75
Polyester	2.00	20–30	0	0.25	150
Silicone	1.88	35–45	30–40	0.4	450

Activity 2.10

A company called DuPont invented Mylar polyester film in the early 1950s. Investigate the development and use of Mylar in the manufacture of electronic components. Write a brief word-processed report giving your findings.

Electronics

The information revolution has largely been made possible by developments in electronics and in the manufacture of integrated circuits in particular. Ongoing improvements in manufacturing technology have given us increasingly powerful integrated circuit chips. Of these, the microprocessor (a chip that performs all of the essential functions of a computer) has been the most notable development. One of these chips is the subject of our case study.

Case study

The microprocessor

A microprocessor is a single chip of silicon that performs all of the essential functions of a computer *central processor unit* (CPU) on a single silicon chip. Microprocessors are found in a huge variety of applications including engine management systems, environmental control systems, domestic appliances, video games, fax machines, photocopiers, etc.

The CPU performs three functions: it controls the system's operation; it performs algebraic and logical operations; and it stores information (or *data*) whilst it is processing. The CPU works in conjunction with other chips, notably those that provide random access memory (RAM), read-only memory (ROM), and input/output (I/O).

The key process in the development of increasingly powerful microprocessor chips is known as *microlithography*. In this process the circuits are designed and laid out using a computer before being

What do each of the following abbreviations stand for?

(a) CPU
(b) RAM
(c) ROM
(d) LSI
(d) VLSI.

photographically reduced to a size where individual circuit lines are about 1/100 the size of a human hair. Early miniaturization techniques, which were referred to as large-scale integration (LSI), resulted in the production of the first generation of 256K-bit memory chip (note that such a chip actually has a storage capacity of 262,144-bits where each bit is a binary 0 or 1). Today, as a result of very-large-scale integration (VLSI), chips can be made that contain more than a million transistors.

Figure 2.13 *A VLSI chip*

Activity 2.11

The die used to produce a microprocessor is cut from a thin wafer of silicon on which a number of identical chips are fabricated. Draw a series of sketches that illustrate the process of chip manufacture, starting with a cylinder of pure silicon and ending with a chip encapsulated in a package with pins that allow it to be connected to a socket mounted on a printed circuit board.

One of the most popular microprocessors to appear in the last 20 years was originally conceived in 1977 by a project team at Motorola. The principal architect of this chip, the 68000, was a man called Tom Gunter. The project was known as 'Motorola Advanced Computer System on Silicon' (MACSS). At the time, Motorola was considering what direction to take in the development of their existing 8-bit microprocessors. Tom Gunter proposed a 16-bit microprocessor extendible to a full 32 bits. At the time, this was a radical departure from the current state-of-the-art, which centred around 8-bit microprocessors employed in systems with a somewhat limited memory capacity (64 Kbytes maximum).

Tom Gunter proved to be a visionary with his proposal for a highly complex microprocessor (containing the equivalent of approximately 68,000 transistors). Gunter's device could only be manufactured in 3 micron (µm) HMOS (a process which had not, at the time, been perfected). It would employ 32-bit data paths and two separate arithmetic logic units. Furthermore, the microprocessor would require a massive 64-pin package (only 40-pin packages had been used at that time) and the die used in the production of the semiconductor chip would have to be very much larger than anything that had ever been used before.

It was something of a testimonial to Motorola's faith in its development team (which included development, software and fabrication engineers) that the project to develop the 68000 went ahead. The concept was to create a new family of super-microprocessors that would provide increasing levels of functionality and performance.

At the time, it was felt that the majority of future software development would be in high-level languages and that this should be reflected in both the internal architecture and instruction set of the 68000. Thus the 68000 chip was given a particularly sleek and uncluttered architecture.

The 68000 employs what has become commonly known as a *complex instruction set*. Computers that employ this technology are known as CISC (complex instruction set computer) machines.

Activity 2.12

Find out about the HMOS process. Explain what it is and how it is used in the production of microprocessors and other VLSI chips. Present your findings in the form of a brief technical report.

Activity 2.13

CISC machines are not the only types of computer. Find out about the alternative RISC technology. Why is it different and how does it compare? Present your findings in the form of a brief article for a local computer club.

Automation

Automation has been widely introduced into the engineering industry and can be defined as the use of mechanical and electrical/electronic systems to carry out processes and functions without requiring direct human control. Automation is often associated with the use of robots but it can also be applied to manufacturing processes that operate under computer (programmed) control. Our case study introduces a particular example of automation in the development of remotely operated vehicles (ROVs).

Case study	Remotely operated vehicles

Remotely operated vehicles (ROV) are designed to work in environments in which humans could not survive. Such environments include the depths of the world's oceans as well as outer space. If you saw a film of the wreck of the Titanic or the recent landing on Mars you will already have experienced the use of the vehicles!

There is no operator or driver actually sitting in an ROV, instead a remote control system is used to operate the vehicle and the 'pilot' remains at some distance from the point at which the vehicle is operating (a very considerable distance in the case of the Mars lander).

Many different companies are involved with the design, manufacture and operation of ROVs. A typical ROV system has at least four major system components. They are:

- the vehicle chassis and drive mechanism
- a deployment system that gets the vehicle to where it is to be used (and also recovers it where possible)
- a remote control system which allows the operator to work remotely from the vehicle
- a cable, fibre or radio telemetry system that links the vehicle with the control console.

ROVs are able to perform many tasks including:

- marine surveying, search and recovery
- deep sea cable burial, inspection, maintenance and repair
- marine engineering, construction and repair
- space and interplanetary exploration
- fire fighting (particularly in the chemical and oil industries)
- inspection, maintenance and repair in the nuclear industry
- bomb disposal.

ROV systems usually require advanced materials and construction methods, and are usually 'one-of-a-kind' vehicles. For example, many deep sea ROVs are designed for operation at a particular depth. ROVs for space exploration, on the other hand, are designed to operate under an extreme range of temperatures and are usually designed for a single mission (they may often not be recoverable).

The typical specification for a deep sea ROV designed for surveying, search and recovery is shown in Table 2.3.

Activity 2.14

Use the Internet or other information sources to locate information and specifications of at least three deep–sea ROVs. Compare these specifications and provide details of typical projects in which these vehicles have been used. Present your findings in the form of a series of word processed fact sheets.

Table 2.3 *Typical specification for a deep sea ROV*

Depth	2 km
Lifting capacity	670 kg (without additional buoyancy)
Payload	225 kg (without additional buoyancy)
Propulsion	Eight hydraulic thrusters (two axial, four vertical and two lateral)
Forward speed	3 knots (1.5 m/s) at 1,000 lbs (4.5 kN) thrust
Hydraulic power	Twin 50 HP (37.5 kW) hydraulic power units
Cameras	Colour camera with pan and tilt plus wide angle black and white camera
Instruments	Sonar, depth, and gyro heading sensors
Illumination	Six 250 W lights
Manipulators	Two 7-function manipulators
Cutters	Abrasive wheel (capable of cutting up to 76 mm diameter cable) plus hydraulic guillotine

Activity 2.15

Use the Internet or other electronic information sources to find out how a gyro works. Present your findings in the form of a brief illustrated article for an engineering club newsletter.

Activity 2.16

Use your school or college library or other paper-based information sources to find out how sonar works. Present your findings in the form of a brief illustrated article for an engineering club newsletter.

Activity 2.17

Sketch a labelled diagram showing the propulsion system used in the ROV described in Table 2.3. The lifting capacity of the ROV is to be increased. Suggest, with the aid of a diagram, a method of providing additional buoyancy. Present your work in the form of a series of overhead projector transparencies.

Test your knowledge 2.12

List the four main system components of an ROV. Explain briefly what each system component is used for.

One of the payloads carried by the recent Mars Pathfinder spacecraft was a miniature ROV called 'Sojourner'. The ROV was deployed on the surface of Mars on 4th July 1997 following a seven-month journey through interplanetary space. After landing the microrover carried out a variety of experiments to determine wheel–soil interactions, navigation and hazard avoidance capabilities. Sojourner also carried an alpha proton X-ray spectrometer (APXS) that was used to determine the element composition of the soil and rocks on Mars.

Sojourner was a six-wheeled vehicle of a rocker bogie design (see Figure 2.14). The ROV weighed a mere 11.5kg (25 lbs) and was about the size of a milk crate. The six-wheeled design allowed the ROV to traverse obstacles of up to a wheel diameter (13 cm) in size. Each wheel was independently actuated and geared (2000:1) providing superior climbing capability in soft sand. The front and rear wheels were independently steerable, providing the capability

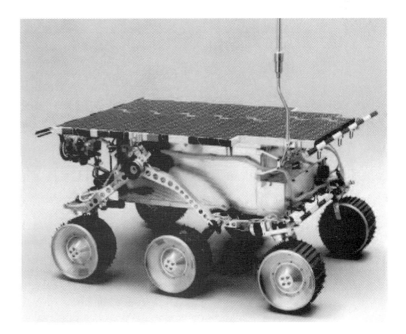

Figure 2.14 *Sojourner*

for the vehicle to turn in place. The vehicle had a top speed of 0.4 m/min.

Sojourner was powered by a 0.22 m^2 solar panel comprising of 13 sets of 18, 5.5 mm Gallium Arsenide (GaAs) solar cells (see Table 2.4). The solar panel was backed up by nine Lithium–Chloride primary batteries, providing up to 150 Whr. This combined panel/batteries system allowed Sojourner's system to demand up to 30 W of peak power. The normal driving power requirement for the microrover was a modest 10 W.

Since the temperature on the surface of Mars can fall to –110°C during the Martian night, some of Sojourner's components were enclosed in a warm 'electronics box' (WEB). The WEB was insulated, coated with high and low emissivity paints, and heated

Table 2.4 *Specification for Sojourner's solar cell battery*

Construction	Gallium arsenide on germanium (GaAs/Ge)
Configuration	13 parallel strings, each with 18 series-connected cells
Operating voltage	14–18 V
Temperature	-140°C to +110°C (max)
Power output	16.5 W on Mars (at noon)
Hydraulic power	Twin 50 HP (37.5 kW) hydraulic power units
Substrate	Nomex honeycomb
Weight	0.34 kg
Area	0.22 m^2
Coverglass	3 mm
Efficiency	>18%

with a combination of three 1 W resistive heater units and waste heat produced by the other electronics components. The design allowed the WEB to keep Sojourner's electronic components at an ambient temperature of between -40°C and +40°C during a complete Martian day.

Control was provided by an 80C85 microprocessor which was capable of executing 100,000 instructions per second. The microprocessor system was fitted with 176 Kbytes of PROM and 576 Kbytes of RAM and 70 I/O channels for sensors and devices such as cameras, modem, motors and experimental electronics.

Vehicle motion control was accomplished through the on/off switching of the drive or steering motors. An average of motor encoder (drive) or potentiometer (steering) readings determined when to switch off the motors. When the motors were off, the microprocessor system was programmed to carry out a proximity and hazard detection function, using its laser and camera system to determine the presence of obstacles in its path. The vehicle was able to avoid obstacles in its path without operator intervention and yet still arrive at its commanded goal location. While stopped, the microprocessor system updated its measurement of distance travelled and heading using the average speed and the on-board gyro. This allowed the microrover to make regular estimates of its progress to the goal location.

Command and telemetry was provided by modulator–demodulators (modems) on the microrover and lander. The microrover was linked to the lander by means of a short–range UHF radio system. During the Martian day, the microrover regularly requested transmission of any commands sent from earth that were stored in the lander's memory. When commands were not available, the microrover transmitted any telemetry collected during the last interval between communication sessions. The telemetry received by the lander from the microrover was stored and forwarded to the earth as part of the lander's own telemetry. In addition, this communication system was used to provide a 'heartbeat' signal during vehicle driving.

When stopped, the microrover sent signals to the lander. Once acknowledged by the lander, the microrover then moved to the next stopping point along its track. Commands for the microrover were sent, via the lander, from the mission's earth control station. In the opposite direction, the lander transmitted its stereo images of the microrover back to the earth. These images, portions of a terrain panorama and supporting images from the microrover's own cameras were displayed at the control station. The operator was then able to designate the display points in the terrain that were to serve as goal locations for the microrover. The coordinates of these points were transferred into a file containing the commands for execution by the microrover and then sent from earth to the lander for onward transmission to the microrover. In addition, the operator was able to use a model which, when overlayed on the image of the vehicle, measured the location and heading of the ROV. This information was also transferred into the command file to be sent to the microrover on its next traverse in order to correct any navigation errors.

Activity 2.18

Use the Internet or other electronic information sources to find out how a modem works. Present your findings in the form of a brief illustrated article for an engineering club newsletter.

Activity 2.19

Use your school or college library, electronic data books, or other paper-based information sources to obtain data on the 80C85 microprocessor (including, as a minimum, absolute maximum ratings and pin connections). Present your findings in the form of an illustrated data sheet.

Activity 2.20

Sketch a labelled diagram showing the communication systems used to:
(a) send commands from the earth control station to Sojourner
(b) receive telemetry data sent from Sojourner at the earth control station.
Present your work in the form of a drawing produced by a CAD or computer drawing package.

Test your knowledge 2.13

Explain how a typical ROV is able to accept commands from a remote location.

Investigating new technology products

The engineering sectors and typical new technology products with which you need to be familiar are as follows:

Aerospace
New passenger and military aircraft, satellites, space vehicles, missiles, etc. from companies such as British Aerospace, Westland and Rolls-Royce.

Electrical and electronic
Electric generators and motors, consumer electronic equipment (radio, TV, audio and video) power cables, computers, etc. produced by companies such as GEC, BICC and ICL.

Mechanical
Bearings, agricultural machinery, gas turbines, machine tools and the like from companies such as RHP, GKN and Rolls-Royce.

Telecommunications
Telephone, radio and data communications equipment, etc. from companies like Nokia, GEC, Plessey and British Telecom.

Within the above sectors you need to be able to identify and investigate a variety of products that are based on the use of new technology. Our final case study provides you with an example of the application of several new technologies that you will already be familiar with!

Activity 2.21

Investigate the engineering industry in your area. Name at least three engineering companies and identify the engineering sectors in which they are active. Give examples of how these companies are making use of new technology. Present your findings in the form of a brief class presentation using appropriate visual aids.

Test your knowledge 2.14

To which sectors do each of the following new technology products belong:

(a) mobile phones
(b) portable CD players
(c) variable speed drives
(d) microprocessors.

Activity 2.22

New materials are being used in many branches of engineering, eventually finding their way into products that we use in the home. Investigate three different everyday products that use new materials. In each case explain how and why the material is used and, where appropriate, name the materials that have been replaced. Present your findings in the form of a set of word processed 'fact sheets'.

| Case study | # Compact disks |

Compact disks can provide 65 minutes of high-quality recorded audio or 650 Mbytes of computer data that is roughly equivalent to 250,000 pages of A4 text. It is therefore hardly surprising that the compact disk has now become firmly established in both the computing and hi-fi sectors.

As with most 'new' products, the technology used in compact disks (and in compact disk players and recorders) relies on several other technologies working together. In the case of compact disks, the most important of these are:

- digital audio technology (being able to represent audio signals using a sequence of digital codes)
- optical technology (being able to produce a precisely focused beam from a laser light source)
- control system technology (being able to control motor speed as well as the position and focus of the optical unit)
- manufacturing technology (being able to reliably and cost-effectively manufacture both the equipment used to play compact disks and the compact disk media itself).

Before explaining how compact disk technology works, it is worth looking at the previous storage technology in which vinyl disks were used for recording analogue signals in the form of 'long-play' (LP) and 'extended-play' (EP) 'records'. Despite its limitations, vinyl disk recording technology survived for nearly four decades (from around 1950 to around 1990). However, the problems associated with recording analogue signal variations in a groove pressed into the relatively soft surface of plastic eventually led to the downfall of the LP recording and its replacement by the digital compact disk in the late 1980s.

A comparison between the performance specifications of a compact disk system with those of an LP record player reveals a number of important differences, as shown in Table 2.5.

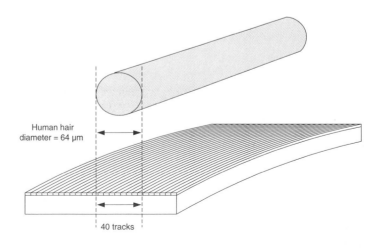

Figure 2.15 *A human hair compared with the tracks on a CD*

Table 2.5 *Comparison of CD and LP record player specifications*

Specification	Compact disk player	LP record player
Recording technique	Digital	Analogue
Material	Glass	Vinyl
Dynamic range	90 dB typical	70 dB typical
Frequency response	20 Hz to 20 kHz ± 0.5 dB	20 Hz to 20 kHz ± 3 dB
Signal-to-noise ratio	90 dB typical	60 dB typical
Harmonic distortion	Less than 0.01%	Less than 2%
Channel separation	More than 90 dB	Less than 40 dB
Wow and flutter	Better than 1 part in 10^5	Better than 0.1%
Diameter	120 mm	305 mm
Rotational velocity	196 to 568 rpm	33.3 rpm
Playing time (per side)	65 minutes typical	20 minutes typical

Conventional CD-ROMs, like audio compact disks, are made up of three basic layers. The main part of the disk consists of an injection-moulded polycarbonate *substrate* which incorporates a spiral track of *pits* and *lands* that are used to encode the data that is stored on the disk. Over the substrate is a thin aluminium (or gold) reflective layer, which in turn is covered by an outer protective lacquer coating.

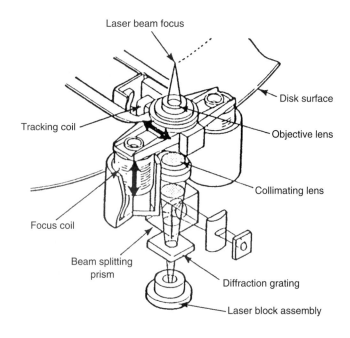

Figure 2.16 *Optical assembly fitted to a CD-ROM drive*

Test your knowledge 2.15

Give THREE reasons why compact disks offer vastly improved performance when compared with LP records.

Test your knowledge 2.16

What do the following abbreviations stand for?

(a) CD
(b) CD-ROM
(c) CD-R
(d) CD-E

Test your knowledge 2.17

Explain why precise focusing of the laser beam in a CD player is essential.

Information is retrieved from the CD by focusing a low-power (0.5 mW) infrared (780 nm) laser beam onto the spiral track of pits and lands in the disk's substrate. The height difference between the pits and the adjacent lands creates a phase shift causing destructive interference in the reflected beam. The effect of the destructive interference and light scattering is that the intensity of the light returned to a photodiode detector is modulated by the digital data stored on the disk. This modulated signal is then processed, used for tracking, focus, and speed control, and then decoded and translated into usable data. The optical assembly, complete with lenses and focus coils, is shown in Figure 2.16.

Conventional compact disks and CD-ROMs only support playback (or reading) of the data stored in them. In recent years new technology has appeared which supports both playback (reading) and recording (writing). This technology has resulted in two new types of compact disk; the CD-R (recordable) and the CD-E (erasable).

Activity 2.23

Investigate the construction of a typical portable CD player. Itemize each of the main components or sub-assemblies used and describe the materials and processes used in its manufacture. Present your findings in the form of a brief class presentation using appropriate visual aids. Comment on any safety issues that you encounter.

Review questions

1 For each of the engineering sectors listed below, identify a product or service which is based on the application of new technology and, in each case, explain briefly how the technology in question has influenced the development of the product:

(a) Information and communications technology
(b) New materials and components
(c) Automation.

2 Explain the following terms used in computer aided engineering (CAE):

(a) computer aided design (CAD)
(b) computer aided manufacture (CAM)
(c) computer integrated manufacturing (CIM)
(d) computer numerical control (CNC).

3 Explain how a database management system (DBMS) is used in a typical engineering company.

4   Explain the following terms in relation to the Internet and the World Wide Web:
(a) Uniform resource locator (URL)
(b) Hypertext transfer protocol (HTTP)
(c) Hypertext markup language (HTML).

5   Explain how an engineering company can use an Intranet to improve communications between employees and departments.

6   With reference to the Internet and the World Wide Web, explain the function of a search engine.

7   With the aid of a diagram, explain how data can be transferred from one computer to another using an optical fibre link.

8   Explain the essential difference between thermoplastics and thermosetting materials (thermosets).

9   State TWO thermoplastic materials and TWO thermosetting materials.

10 Categorize each of the following materials as either a thermoplastic material or a thermosetting material:

(a) Polythene
(b) Polyester
(c) Silicone
(d) Teflon.

11   State a typical application for each material in Question 10.

12   Describe THREE applications of microprocessors.

13   Explain, with the aid of a simple diagram, how a GPS receiver is able to determine its position accurately on the surface of the earth.

14   Describe TWO applications of ROVs.

15   Explain how data is stored on the surface of a compact disk.

16   Identify THREE different uses of new technology within a compact disk player.

Unit 3 Make engineered products

Summary

This unit introduces you to a selection of widely used engineering materials and components. It also introduces you to a range of workshop and manufacturing processes. Some of these processes you will eventually use in your project work. Others are more appropriate to the large scale production of materials and components. We begin by introducing two engineered products with which you may already be familiar—an electrician's screwdriver and a bench power supply. This unit has links with many other Intermediate units and in particular with Unit 1 (Design and graphical communication) and Unit 2 (Application of new technology).

Products

Engineered products are usually assemblies of individual components. Such products can be divided into three main groups:

- Mechanical products such as gear boxes, pumps, engines and turbines.
- Electronic products such as the completed circuit boards used in computers, video recorders and television receivers.
- Electromechanical devices can range from washing machines to computer controlled machine tools. These are devices which combine electrical and mechanical components.

An electrician's screwdriver

Figure 3.1 shows an exploded view of a typical electrician's screwdriver. It has to combine the functions of a conventional screwdriver with the ability to indicate whether a mains circuit is live. It must be strong enough to tighten and undo the small brass

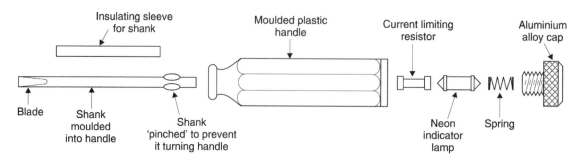

Figure 3.1 *A typical electrician's screwdriver*

screws found in the terminals of electrical accessories. It must be insulated to withstand the potentials (voltages) met with in domestic, industrial and commercial installations. The current through the neon indicator lamp must be limited to a safe level under all conditions. It must be light in weight, compact and competitively priced.

These criteria can be met by careful selection of materials and manufacturing processes.

* The blade is made from a toughened medium carbon steel (0.8% carbon).
* The shank of the blade is insulated using a PVC sleeve.
* The handle is moulded from cellulose acetate. This is a tough, flame resistant plastic with good insulating properties. It is transparent so that the neon indicator lamp can be seen to light up. The blade would be moulded into the handle.
* The spring would be made from hard drawn phosphor bronze wire. This is a good conductor and corrosion resistant.
* The end cap would be made from an aluminium alloy on a computer numerically controlled (CNC) lathe. This material is light in weight, easily cut, a good conductor and corrosion resistant. The process of manufacture is suitable for large batch production.
* The neon indicator lamp and the current limiting resistor would be bought in as standard components.

All the materials chosen are readily available and relatively low in cost. They are selected for their fitness for purpose. Such screwdrivers are made and sold in large quantities and the manufacturing processes chosen lend themselves to large batch production.

A bench power supply

Figure 3.2 shows the circuit for a variable voltage d.c. power supply unit. It also shows some of the constructional details.

Most of the components for this project such as the transformer, rectifier pack, the sockets, switches, fuse holders, etc. would be bought in. The charger will have to have quite a high current output, so you must carefully consider how to keep the components cool. This particularly applies to the rectifier pack since solid state devices

CIRCUIT DIAGRAM

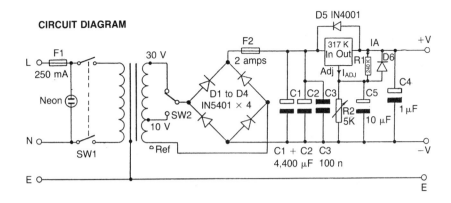

CONSTRUCTION

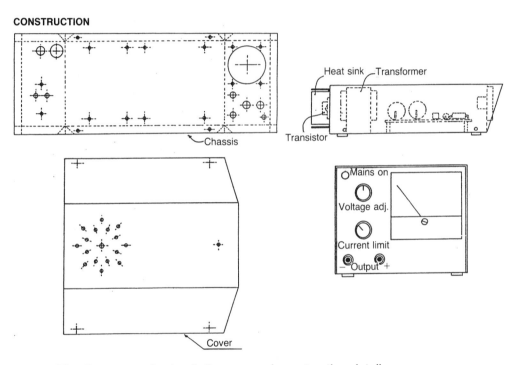

Figure 3.2 *The Power supply circuit diagram and construction details*

are destroyed if they are allowed to overheat. For this reason an aluminium case should be used and a heatsink if recommended by the manufacturer of the rectifier. Cooling holes or louvres should be included in your design. These should be positioned so that moisture cannot enter the case. Since the equipment is connected to the mains supply and the case is of metal construction, the case must be earthed in accordance with current safety regulations.

Product specifications

In order to make an engineered product you need to understand the requirements of the product specification and be able to use the information contained in the specification to make decisions about the development and manufacture of the product.

A product specification is simply a technical description of the key features of the product such as size (i.e. dimensions), weight, materials, finish, and performance. Specifications need to be agreed (and understood) before any attempt is made to plan the manufacture of a product. Failure to do this may mean that the finished product fails to meet the specification and therefore does not satisfy the needs of the customer or 'end user'. You should refer back to Page 79 for a full 'designer's checklist' and Activity 1.13 which required you to produce a specification for a simple consumer electronic product.

Activity 3.1

Obtain the detailed specification for a typical domestic hi-fi amplifier. Write a brief article for a consumer magazine explaining what each of the specifications means and how they should be interpreted by a potential purchaser. Present your work in the form of a word processed document.

Production planning

Having developed a detailed specification (and agreed this with the customer or client), the next stage is that of preparing a *production plan*. The production plan provides all of the details needed to actually make the product. The plan should not be confused with the product specification—the former contains information about the processes used and the sequence of assembly whilst the latter relates to the performance, appearance, quality and styling of the finished product.

A typical production plan will contain details of:
- materials to be used
- processes to be used
- sequence of assembly/manufacture
- equipment and resources required for assembly/manufacture
- scheduling and timing of assembly/manufacture
- testing and quality checks.

Finally, it is worth noting that individual production plans will depend very much on the nature of the engineered product to which they relate.

Engineering materials

To make the engineering products that have just been described, you had to select suitable materials and components. We are now going to consider some of the main groups of materials and components available to engineers.

Figure 3.3 shows the main groups of engineering materials. The Latin name for iron is ferrum, so it is not surprising that ferrous

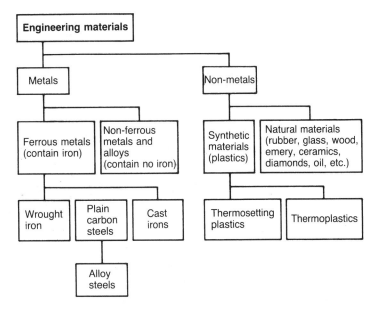

Figure 3.3 *The main groups of engineering materials*

metals and alloys are all based on the metal iron. Alloys consist of two or more metals or metals and non-metals that have been brought together as compounds or solid solutions to produce a metallic material with special properties. For example an alloy of iron, carbon, nickel and chromium is stainless steel. This is a corrosion resistant ferrous alloy. The remainder of the metallic materials available are non-ferrous metals and alloys. Non-metals can be natural, such as rubber, or they can be synthetic such as the plastic compound PVC (see Page 99).

Properties of materials

When selecting a material we need to make certain that it has suitable properties for the job it has to do. For example we must ask ourselves the following questions.

- Will it corrode in its working environment?
- Will it weaken or melt in a hot environment?
- Will it break under normal working conditions?
- Can it be easily cast, formed or cut to shape?

We compare materials by comparing their properties. We will now consider the more important properties of engineering materials.

Chemical properties

Corrosion

This is caused by the metals and metal alloys being attacked and eaten away by chemical substances. For example the rusting of ferrous metals and alloys is caused by the action of atmospheric oxygen in the presence of water. Another example is the attack on aluminium and some of its alloys by strong alkali solutions. Take care when using degreasing agents on such metals. Copper and copper based alloys are stained and corroded by the active sulphur and chlorine products found in some heavy duty cutting lubricants. Choose your cutting lubricant with care when machining such materials.

Degradation

Non-metallic materials do not corrode but they can be attacked by chemical substances. Since this weakens or even destroys the material it is referred to as degradation. Unless specially compounded, rubber is attacked by prolonged exposure to oil. Synthetic (plastic) materials can be softened by chemical solvents. Exposure to the ultraviolet rays of sunlight can weaken (perish) rubbers and plastics unless they contain compounds that filter out such rays.

Physical properties

Electrical resistance

Materials with a very low resistance to the flow of an electric current are good electrical conductors. Materials with a very high resistance to the flow of electric current are good insulators. Generally, metals are good conductors and non-metals are good insulators (poor conductors). A notable exception is carbon which conducts electricity despite being a non-metal. You will learn about electrical resistance in your science unit. The electrical resistance of a conductor depends upon:

- its length (the longer it is the greater its resistance)
- its thickness (the thicker it is the lower its resistance)
- its temperature (the higher its temperature the greater is its resistance)
- its resistivity (this is the resistance of a material measured between the opposite faces of a metre cube of the material).

A small number of non-metallic materials, such as silicon, have atomic structures that fall between electrical conductors and insulators. These materials are called semi-conductors and are used for making solid state devices such as transistors.

Magnetic properties

All materials respond to strong magnetic fields to some extent. Only the ferro-magnetic materials respond sufficiently to be of interest. The more important of the ferro-magnetic materials are the metals iron, nickel and cobalt. Soft magnetic materials, such as soft iron, can be magnetized by placing them in a magnetic field. They cease to be magnetized as soon as the field is removed.

Hard magnetic materials, such as high-carbon steel that has been hardened by cooling it rapidly (quenching) from red heat, also become magnetized when placed in a magnetic field. Hard magnetic materials retain their magnetism when the field is removed. They become permanent magnets.

Permanent magnets can be made more powerful for a given size by adding cobalt to the steel to make an alloy. Soft magnetic materials can be made more efficient by adding silicon or nickel to the pure iron. Silicon–iron alloys are used for the rotor and stator cores of electric motors and generators. Silicon–iron alloys are also used for the cores of power transformers.

Thermal properties

These include:

- The melting temperatures of materials. Note that some plastics do not soften when heated, they only become charred and are destroyed. This will be considered later in this unit.
- Thermal conductivity. This is the ease with which materials conduct heat. Metals are good conductors of heat. Non-metals are poor conductors of heat. Therefore non-metals are heat insulators.
- Expansion. Metals expand appreciably when heated and contract again when cooled. They have high coefficients of linear expansion. Non-metals expand to a lesser extent when heated. They have low coefficients of linear expansion. Again, these properties will be considered in more detail in your science unit.

Mechanical properties

Strength

This is the ability of a material to resist an applied force (load) without fracturing (breaking). It is also the ability of a material not to yield. Yielding is when the material 'gives' suddenly under load and changes shape permanently but does not break. This is what happens when metal is bent or folded to shape. The load or force can be applied in various ways, as shown in Figure 3.4.

You must be careful when interpreting the strength data quoted for various materials. A material may appear to be strong when subjected to a static load, but will break when subjected to an impact load. Materials also show different strength characteristics when the load is applied quickly from when the load is applied slowly.

Toughness

This is the ability of a material to resist impact loads as shown in Figure 3.5. Here, the toughness of a piece of high-carbon steel in the soft (annealed) condition is compared with a piece of the same steel after it has been hardened by raising it to red-heat and cooling it quickly (quenching it in cold water). The hardened steel shows a greater strength, but it lacks toughness.

A test for toughness, called the Izod test, uses a notched specimen that is hit by a heavy pendulum. The test conditions are carefully controlled, and the energy absorbed in bending or breaking the specimen is a measure of the toughness of the material from which it was made.

Elasticity

Materials that change shape when subjected to an applied force but spring back to their original size and shape when that force is removed are said to be elastic. They have the property of *elasticity*.

Test your knowledge 3.1

List THREE main factors that must be considered when selecting a material for a given component.

Test your knowledge 3.2

Give two examples of corrosion in metals or metal alloys.

Test your knowledge 3.3

On what does the resistance of an electrical conductor depend?

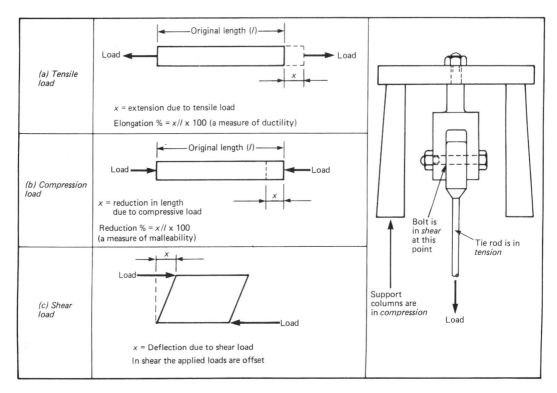

Figure 3.4 *Different ways in which a load can be applied*

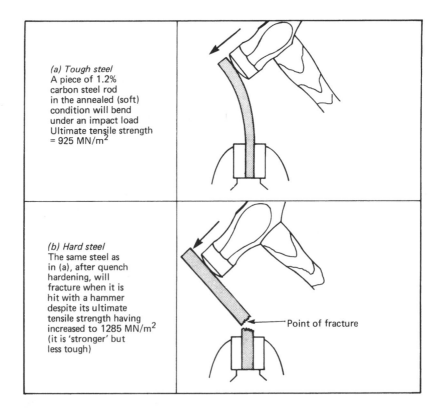

Figure 3.5 *Impact loads*

Plasticity

Materials that flow to a new shape when subjected to an applied force and keep that shape when the applied force is removed are said to be plastic. They have the property of *plasticity*.

Ductility

Materials that can change shape by plastic flow when they are subjected to a pulling (tensile) force are said to be ductile. They have the property of *ductility* (see Figure 3.6).

Malleability

Materials that can change shape by plastic flow when they are subjected to a squeezing (compressive) force are said to be malleable. They have the property of *malleability* (see Figure 3.6).

Hardness

Materials that can withstand scratching or indentation by an even harder object are said to be hard. They have the property of hardness. Figure 3.7 shows the effect of pressing a hard steel ball into two pieces of metal with the same force. The ball sinks further in to the softer of the two pieces of metal than it does into the harder metal.

There are various hardness tests available. The Brinell hardness test uses the principles set out above. A hardened steel ball is pressed into the specimen by a controlled load. The diameter of the indentation is measured using a special microscope. The hardness

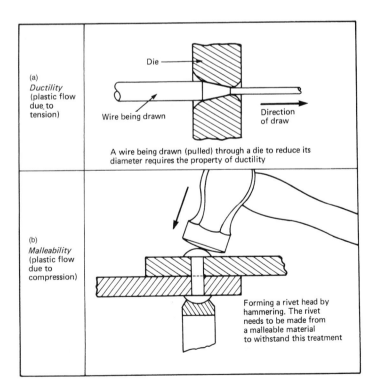

Figure 3.6 *Two common engineering processes, drawing and riveting. Drawing exploits ductility whilst riveting exploits malleability*

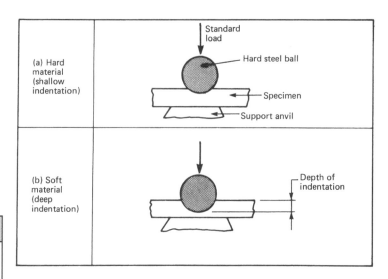

Figure 3.7 *The effect of pressing a hard steel ball into two materials with different hardness properties.*

number is obtained from the measured diameter by use of conversion tables.

The Vickers test is similar but uses a diamond pyramid instead of a hard steel ball. This enables harder materials to be tested. The diamond pyramid leaves a square indentation and the diagonal distance across the square is measured. Again, conversion tables are used to obtain the hardness number from the measured distance.

The Rockwell test uses a diamond cone. A minor load is applied and a small indentation is made. A major load is then added and the indentation increases in depth. This increase in depth of the indentation is directly converted into the hardness number and it can be read from a dial on the machine.

Rigidity

Materials that resist changing shape under load are said to be rigid. They have the property of rigidity. The opposite of rigidity is flexibility. Rigid materials are usually less strong than flexible materials. For example, cast iron is more rigid than steel but steel is the stronger and tougher. However the rigidity of cast iron makes it a useful material for machine frames and beds. If such components were made from a more flexible material the machine would lack accuracy. It would be deflected by the cutting forces.

Ferrous metals

As previously stated, ferrous metals are based upon the metal iron. For engineering purposes iron is usually associated with various amounts of the non-metal carbon. When the amount of carbon present is less than 1.8% we call the material steel. The figure of 1.8% is the theoretical maximum. In practice there is no advantage in increasing the amount of carbon present above 1.4%. We are only going to consider the plain carbon steels. Alloy steels are beyond the scope of this book. The effects of the carbon content on the properties of plain carbon steels are shown in Figure 3.8.

Cast irons are also ferrous metals. They have substantially more carbon than the plain carbon steels. Grey cast irons usually have a carbon content between 3.2% and 3.5%. Not all this carbon can be taken up by the iron and some is left over as flakes of graphite between the crystals of metal. It is these flakes of graphite that gives cast iron its particular properties and makes it a 'dirty' metal to machine. The compositions and typical uses of plain carbon steels and a grey cast iron are summarized in Table 3.1.

Low carbon steels

These are also called mild steels. They are the cheapest and most widely used group of steels. Although they are the weakest of the steels, nevertheless they are stronger than most of the non-ferrous metals and alloys. They can be hot and cold worked and machined with ease.

Medium carbon steels

These are harder, tougher, stronger and more costly than the low carbon steels. They are less ductile than the low carbon steels and cannot be bent or formed to any great extent in the cold condition without risk of cracking. Greater force is required to bend and form

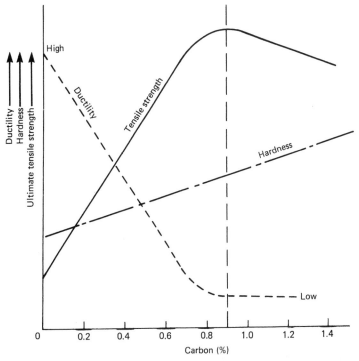

Figure 3.8 *Effect of carbon content on the properties of plain carbon steels*

Table 3.1 *Ferrous metals*

Name	Group	Carbon content (%)	Some uses
Dead mild steel (low carbon steel)	Plain carbon steel	0.10–0.15	Sheet for pressing out components such as motor car body panels. General sheet-metal work. Thin wire, rod and drawn tubes.
Mild steel (low carbon steel)	Plain carbon steel	0.15–0.30	General purpose workshop rod, bars and sections. Boiler plate. Rolled steel beams, joists, angles, etc.
Medium carbon steel	Plain carbon steel	0.30–0.50	Crankshafts, forgings, axles, and other stressed components.
		0.50–0.60	Leaf springs, hammer heads, cold chisels, etc.
High carbon steel	Plain carbon steel	0.8–1.0 1.0–1.2 1.2–1.4	Coil springs, wood chisels. Files, drills, taps and dies. Fine-edge tools (knives, etc.)
Grey cast iron	Cast iron	3.2–3.5	Machine castings.

them. Medium carbon steels hot forge well but close temperature control is essential. Two carbon ranges are shown. The lower carbon range can only be toughened by heating and quenching (cooling quickly by dipping in water). They cannot be hardened. The higher carbon range can be hardened and tempered by heating and quenching.

High carbon steels

These are harder, stronger and more costly than medium carbon steels. They are also less tough. High carbon steels are available as hot rolled bars and forgings. Cold drawn high carbon steel wire (piano wire) is available in a limited range of sizes. Centreless ground high carbon steel rods (silver steel) are available in a wide range of diameters (inch and metric sizes) in lengths of 333 mm, 1 m and 2 m. High carbon steels can only be bent cold to a limited extent before cracking. They are mostly used for making cutting tools such as files, knives and carpenters' tools.

Non-ferrous metals and alloys

Non-ferrous metals (i.e. metals that are *not* based on iron) include metals such as aluminium and zinc as well as alloys such as brass and bronze. We shall start by looking at copper—a material that is widely used in electrical engineering.

Test your knowledge 3.7

Select a suitable ferrous metal, and state its carbon content, for each of the following objects.

(a) A hexagon head bolt produced on a lathe
(b) a car engine cylinder block
(c) a cold chisel
(d) a wood-carving chisel
(e) a pressed steel car body panel.

Copper

Pure copper is widely used for electrical conductors and switchgear components. It is second only to silver in conductivity but it is much more plentiful and very much less costly. Pure copper is too soft and ductile for most mechanical applications.

For general purpose applications such as roofing, chemical plant, decorative metal work and copper-smithing, tough-pitch copper is used. This contains some copper oxide which makes it stronger, more rigid and less likely to tear when being machined. Because it is not so highly refined, it is less expensive than high conductivity copper.

There are many other grades of copper for special applications. Copper is also the basis of many important alloys such as brass and bronze, and we will be considering these next. The general properties of copper are:

* relatively high strength
* very ductile so that it is usually cold worked. An annealed (softened) copper wire can be stretched to nearly twice its length before it snaps
* corrosion resistant
* second only to silver as a conductor of heat and electricity
* easily joined by soldering and brazing. For welding, a phosphorous deoxidized grade of copper must be used.

Copper is available as cold-drawn rods, wires and tubes. It is also available as cold-rolled sheet, strip and plate. Hot worked copper is available as extruded sections and hot stampings. It can also be cast. Copper powders are used for making sintered components. It is one of the few pure metals of use to the engineer as a structural material.

Brass

Brass is an alloy of copper and zinc. The properties of a brass alloy and the applications for which you can use it depends upon the amount of zinc present. Most brasses are attacked by sea water. The salt water eats away the zinc (dezincification) and leaves a weak, porous, spongy mass of copper. To prevent this happening, a small amount of tin is added to the alloy. There are two types of brass that can be used at sea or on land near the sea. These are Naval brass and Admiralty brass.

Brass is a difficult metal to cast and brass castings tend to be coarse grained and porous. Brass depends upon hot rolling from cast ingots, followed by cold rolling or drawing to give it its mechanical strength. It can also be hot extruded and plumbing fittings are made by hot stamping. Brass machines to a better finish than copper as it is more rigid and less ductile than that metal. Table 3.2 lists some typical brasses, together with their compositions, properties and applications.

Tin bronze

As the name implies, the tin bronzes are alloys of copper and tin. These alloys also have to have a deoxidizing element present to

Table 3.2 *Properties and applications of brass alloys*

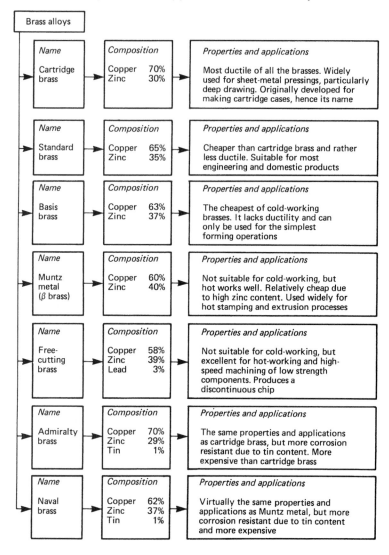

prevent the tin from oxidizing during casting and hot working. If the tin oxidizes the metal becomes hard and 'scratchy' and is weakened. The two deoxidizing elements commonly used are:

- zinc in the gun-metal alloys.
- phosphorus in the phosphor–bronze alloys.

Unlike the brass alloys, the bronze alloys are usually used as castings. However low-tin content phosphor–bronze alloys can be extensively cold worked. Tin–bronze alloys are extremely resistant to corrosion and wear and are used for high pressure valve bodies and heavy duty bearings. Table 3.3 lists some typical bronze alloys together with their compositions, properties and applications.

Aluminium

Aluminium has a density approximately one third that of steel. However it is also very much weaker so its strength/weight ratio is

Table 3.3 *Properties and applications of bronze alloys*

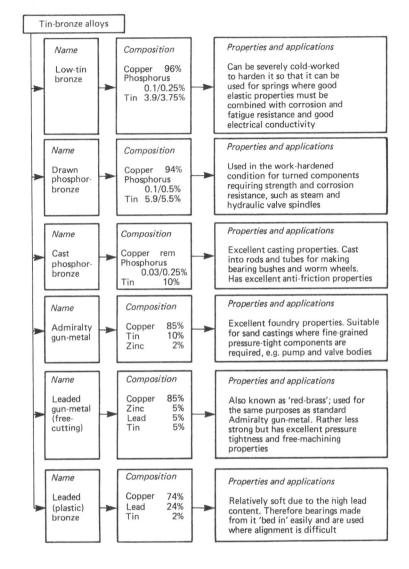

Tin-bronze alloys

Name	Composition	Properties and applications
Low-tin bronze	Copper 96% Phosphorus 0.1/0.25% Tin 3.9/3.75%	Can be severely cold-worked to harden it so that it can be used for springs where good elastic properties must be combined with corrosion and fatigue resistance and good electrical conductivity
Drawn phosphor-bronze	Copper 94% Phosphorus 0.1/0.5% Tin 5.9/5.5%	Used in the work-hardened condition for turned components requiring strength and corrosion resistance, such as steam and hydraulic valve spindles
Cast phosphor-bronze	Copper rem Phosphorus 0.03/0.25% Tin 10%	Excellent casting properties. Cast into rods and tubes for making bearing bushes and worm wheels. Has excellent anti-friction properties
Admiralty gun-metal	Copper 85% Tin 10% Zinc 2%	Excellent foundry properties. Suitable for sand castings where fine-grained pressure-tight components are required, e.g. pump and valve bodies
Leaded gun-metal (free-cutting)	Copper 85% Zinc 5% Lead 5% Tin 5%	Also known as 'red-brass'; used for the same purposes as standard Admiralty gun-metal. Rather less strong but has excellent pressure tightness and free-machining properties
Leaded (plastic) bronze	Copper 74% Lead 24% Tin 2%	Relatively soft due to the high lead content. Therefore bearings made from it 'bed in' easily and are used where alignment is difficult

inferior. For stressed components, such as those found in aircraft, aluminium alloys have to be used. These can be as strong as steel and nearly as light as pure aluminium.

High purity aluminium is second only to copper as a conductor of heat and electricity. It is very difficult to join by welding or soldering and aluminium conductors are often terminated by crimping. Despite these difficulties, it is increasingly used for electrical conductors where its light weight and low cost compared with copper is an advantage. Pure aluminium is resistant to normal atmospheric corrosion but it is unsuitable for marine environments. It is available as wire, rod, cold-rolled sheet and extruded sections for heat sinks.

Commercially pure aluminium is not as pure as high purity aluminium and it also contains up to 1% silicon to improve its strength and stiffness. As a result it is not such a good conductor of electricity nor is it so corrosion resistant. It is available as wire, rod, cold-rolled sheet and extruded sections. It is also available as castings

Select a suitable non-ferrous metal for each of the following objects. Give reasons for your choice.

(a) The body casting of a water pump
(b) screws for clamping the electric cables in the terminals of a domestic electric light switch
(c) a bearing bush
(d) a deep drawn, cup-shaped component for use on land
(e) a ship's fitting made by hot stamping.

and forgings. Being stiffer than high purity aluminium it machines better with less tendency to tear. It forms non-toxic oxides on its surface which makes it suitable for food processing plant and utensils. It is also used for forged and die-cast small machine parts. Because of their range and complexity, the light alloys based upon aluminium are beyond the scope of this unit.

Non-metallic materials

Non-metallic materials can be grouped under the headings shown in Figure 3.9. In addition, wood is also used for making the patterns which, in turn, are used in producing moulds for castings. We are only going to consider some ceramics, thermosets and thermoplastics.

Ceramics

The word ceramic comes from a Greek word meaning potter's clay. Originally, ceramics referred to objects made from potter's clay. Nowadays, ceramic technology has developed a range of materials far beyond the traditional concepts of the potter's art. These include:

• glass products
• abrasive and cutting tool materials
• construction industry materials
• electrical insulators
• cements and plasters for investment moulding
• refractory (heat resistant) lining for furnaces
• refractory coatings for metals

The four main groups of ceramics and some typical applications are summarized in Table 3.4. The common properties of ceramic materials can be summarized as follows:

Strength
Ceramic materials are reasonably strong in compression, but tend to be weak in tension and shear. They are brittle and lack ductility. They also suffer from micro-cracks which occur during the firing process. These lead to fatigue failure. Many ceramics retain their high compressive strength at very high temperatures.

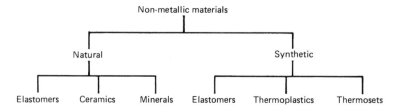

Figure 3.9 *Non-metallic materials*

Table 3.4 *Applications of ceramic materials*

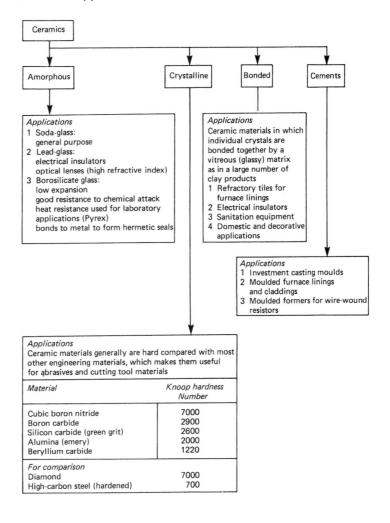

Hardness

Most ceramic materials are harder than other engineering materials, as shown in Table 3.4. They are widely used for cutting tool tips and abrasives. They retain their hardness at very high temperatures that would destroy high carbon and high speed steels. However they have to be handled carefully because of their brittleness.

Refractoriness

This is the ability of a material to withstand high temperatures without softening and deforming under normal service conditions. Some refractories such as high-alumina brick and fireclays tend to soften gradually and may collapse at temperatures well below their fusion (melting) temperatures. Refractories made from clays containing a high proportion of silica to alumina are most widely used for furnace linings.

Electrical properties

As well as being used for weather resistant high-voltage insulators for overhead cables and sub-station equipment, ceramics are now being used for low-loss high-frequency insulators. For example they are being used for the dielectric in silvered ceramic capacitors for high frequency applications.

In all the previous examples the ceramic material is polycrystalline. That is, the material is made up of a lot of very tiny crystals. For solid state electronic devices single crystals of silicon are grown under very carefully controlled conditions. The single crystal can range from 50 mm diameter to 150 mm diameter with a length ranging from 500 mm to 2,500 mm. These crystals are without impurities. They are then cut up into thin wafers and made into such devices as thermistors, diodes, transistors and integrated circuits. This is done by doping the pure silicon wafers with small, controlled amounts of carefully selected impurities. Some impurities give the silicon n-type characteristics. That is they make the silicon electrically negative by increasing the number of electrons present. Some impurities give the silicon p-type characteristics. That is they make the silicon electrically positive by reducing the number of electrons present.

Thermosetting plastics

Themosetting plastics are also known as thermosets. These materials are available in powder or granular form and consist of a synthetic resin mixed with a 'filler'. The filler reduces the cost and modifies the properties of the material. A colouring agent and a lubricant are also added. The lubricant helps the plasticized moulding material to flow into the fine detail of the mould.

The moulding material is subjected to heat and pressure in the moulds during the moulding process. The hot moulds not only plasticize the moulding material so that it flows into all the detail of moulds, the heat also causes a chemical change in the material. This chemical change is called polymerization or, more simply, 'curing'. Once cured, the moulding is hard and rigid. It can never again be softened by heating. If made hot enough it will just burn. Some thermosets and typical applications are summarized in Table 3.5.

Thermoplastics

Unlike the thermosets we have just considered, thermoplastics soften every time they are heated. In fact, any material trimmed from the mouldings can be ground up and recycled. They tend to be less rigid but tougher and more 'rubbery' than the thermosetting materials. Some thermoplastics and typical applications are summarized in Table 3.6.

Reinforced plastics

The strength of plastics can be increased by reinforcing them with fibrous materials.

- *Laminated plastics* (*Tufnol*). Fibrous material such as paper, woven cloth, woven glass fibre, etc. is impregnated with a thermosetting resin. The sheets of impregnated material are laid up in powerful hydraulic presses and they are heated and

Table 3.5 *Thermosetting plastics*

Type	Applications
Phenolic resins and powders	The original 'Bakelite' type of plastic materials, hard, strong and rigid. Moulded easily and heat 'cured' in the mould. Unfortunately, they darken during processing and are only available in the darker and stronger colours. Phenolic resins are used as the 'adhesive' in making plywoods and laminated plastic materials (Tufnol)
Amino (containing nitrogen) resins and powders	The basic resin is colourless and can be coloured as required. Can be strengthened by paper-pulp fillers and are suitable for thin sections. Used widely in domestic electrical switchgear
Polyester resins	Polyester chains can be cross-linked by adding monomer such as styrene, when the polyester ceases to behave as a thermoplastic and becomes a thermoset. Curing takes place by internal heating due to chemical reaction and not by heating the mould. Used largely as the bond in the production of glass fibre mouldings.
Epoxy resins	The strongest of the plastic materials used widely as adhesives, can be 'cold cast' to form electrical insulators and used also for potting and encapsulating electrical components.

squeezed until they become solid and rigid sheets, rods, tubes, etc. This material has a high strength and good electrical properties. It can be machined with ordinary metal working tools and machines. Tufnol is used for making insulators, gears and bearing bushes.

- *Glass reinforced plastics (GRP).* Woven glass fibre and chopped strand mat can be bonded together by polyester or by epoxy resins to form mouldings. These may range from simple objects such as crash helmets to complex hulls for ocean-going racing yachts. The thermosetting plastics used are cured by chemical action at room temperature and a press is not required. The glass fibre is laid up over plaster or wooden moulds and coated with the resin which is well worked into the reinforcing material. Several layers or 'plies' may be built up according to the strength required. When cured the moulding is removed from the mould. The mould can be used again. Note that the mould is coated with a release agent before moulding commences.

Although the properties of plastic materials can vary widely, they all have some general properties in common.

General properties of plastics

Strength/weight ratio
Plastic materials vary considerably in strength and some of the stronger (such as nylon) compare favourably with the metals. All

Table 3.6 *Thermoplastic materials*

Type	Material	Characteristics
Acrylics	Polymethyl-methacrylate	Materials of the 'Perspex' or 'Plexiglass' types. Excellent light transmission and optical properties, tough, non-splintering and can be easily heat-bent and shaped. Excellent high-frequency electrical insulators.
Cellulose plastics	Nitro-cellulose	Materials of the 'celluloid' type. Tough, waterproof, and available as preformed sections, sheets and films. Difficult to mould because of their high flammability. In powder form nitro-cellulose is explosive.
	Cellulose acetate	Far less flammable than nitro-cellulose and the basis of photographic 'safety' film. Frequently used for moulded handles for tools and electrical insulators.
Fluorine plastics (Teflon)	Polytetrafluoro-ethylene (PTFE)	A very expensive plastic material, more heat resistant than any other plastic. Also has the lowest coefficient of friction. PTFE is used for heat-resistant and anti-friction coatings. Can be moulded (with difficulty) to produce components with a waxy feel and appearance.
Nylon	Polyamide	Used as a fibre or as a wax-like moulding material. Tough, with a low coefficient of friction. Cheaper than PTFE but loses its strength rapidly when raised above ambient temperature. Absorbs moisture readily, making it dimensionally unstable and a poor electrical insulator.
Polyesters (Terylene)	Polyethylene-teraphthalate	Available as a film or in fibre form. Ropes made from polyesters are light and strong and have more 'give' than nylon ropes. The film makes an excellent electrical insulator.
Vinyl plastics	Polythene	A simple material, relatively weak but easy to mould, and a good electrical insulator. Used also as a waterproof membrane in the building industry.
	Polypropylene	A more complicated materials than polythene. Can be moulded easily and is similar to nylon in many respects. Its strength lies between polythene and nylon. Cheaper than nylon and does not absorb water.
	Polystyrene	Cheap and can be easily moulded. Good strength but tends to be rigid and brittle. Good electrical insulation properties but tends to craze and yellow with age.
	Polyvinylchloride (PVC)	Tough, rubbery, practically non-flammable, cheap and easily manipulated. Good electrical properties and used widely as an insulator for flexible and semi-flexible cables.

Select a suitable plastic material for the following applications. Give reasons for your choice.

(a) The moulded cockpit cover for a light aircraft
(b) a moulded bearing for a computer printer
(c) a rope for rock climbing
(d) the body of a domestic light switch
(e) encapsulating (potting) a small transformer.

plastics have a lower density than metals and, therefore, chosen with care and proportioned correctly their strength/weight ratio compares favourably with the light alloys.

Corrosion resistance

Plastic materials are inert to most inorganic chemicals and some are inert to all solvents. Thus they can be used in environments that are hostile to the most corrosion resistant metals and many naturally occurring non-metals.

Electrical resistance

All plastic materials are good electrical insulators, particularly at high frequencies. However their usefulness is limited by their softness and low heat resistance compared with ceramics. Flexible plastics such as PVC are useful for the insulation and sheathing of electric cables.

Activity 3.2

Examine and dismantle a 13 A mains plug. Sketch the components for identification purposes (detail not required) and choose a suitable material for each component giving reasons for your choice. Present your work in the form of a one-page 'fact sheet'.

Activity 3.3

Examine and dismantle a typical garden lawn sprinkler of the rotary type. Sketch the components for identification purposes (detail not required) and choose a suitable material for each component giving reasons for your choice. Present your work in the form of a one-page 'fact sheet'.

Activity 3.4

An electrical equipment manufacturer has asked you to advise them on the selection of materials to be used in the manufacture of an adjustable desk lamp which is to be fitted with a switch and is to accept a conventional 40 W light bulb. Sketch a suitable design and list each of the component parts required. Suggest, with reasons, a material to be used for each component. Present your work in the form of an illustrated word processed report.

Engineering components

Wherever possible, rather than manufacture from raw materials, you should use standard commercially available components. Because these are mass produced in very large quantities, their cost is kept to a minimum. If they are made to an internationally acceptable standard, their quality is guaranteed. Let's now take a look at some of the more widely used mechanical and electrical components:

Mechanical components

Screwed fastenings

Screwed fastenings refer to nuts, bolts, screws and studs. These come in a wide variety of sizes and types of screw thread. When selecting a screwed fastening for any particular purpose, you should ask yourself the following questions.

- Is the fastening strong enough for the application?
- Is the material from which the fastening is made corrosion resistant under service conditions and is it compatible with the metals being joined?
- Is the screw thread chosen, suitable for the job? Coarse threads are stronger than fine threads, particularly in soft metals such as aluminium. Fine threads are less likely to work loose.

Figure 3.10 shows some typical screwed fastenings and it also shows how they are used.

There are a large variety of heads for screwed fastenings, and the selection is usually a compromise between strength, appearance and ease of tightening. The hexagon head is usually selected for general engineering applications. The more expensive cap-head screw is widely used in the manufacture of machine tools, jigs and fixtures, and other highly stressed applications. These fastenings are forged from high-tensile alloy steels, thread rolled and heat treated. By recessing the cap-head, a flush surface is provided for safety and easy cleaning. Figure 3.11 shows some alternative screw heads.

Riveted joints

Figure 3.12 shows some typical riveted joints. Riveted joints are very strong providing they are correctly designed and assembled. The joint must be designed so that the rivet is in shear and not in tension. Consider the head of the rivet as only being strong enough to keep the rivet in place. You must consider a number of factors when selecting a rivet and making a riveted joint. These include the material used for the rivet as well as the shape of its head.

The material from which the rivet is made must not react with the components being joined as this will cause corrosion and weakening. Also the rivet must be strong enough to resist the loads imposed upon it.

The rivet head chosen is always a compromise between strength and appearance. In the case of aircraft components, wind resistance must also be taken into account. Figure 3.13 shows some typical rivet heads and rivet types.

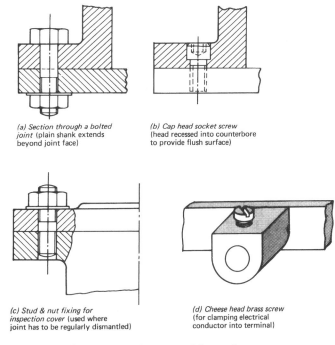

(a) Section through a bolted joint (plain shank extends beyond joint face)

(b) Cap head socket screw (head recessed into counterbore to provide flush surface)

(c) Stud & nut fixing for inspection cover (used where joint has to be regularly dismantled)

(d) Cheese head brass screw (for clamping electrical conductor into terminal)

Figure 3.10 *Some typical screwed fastenings*

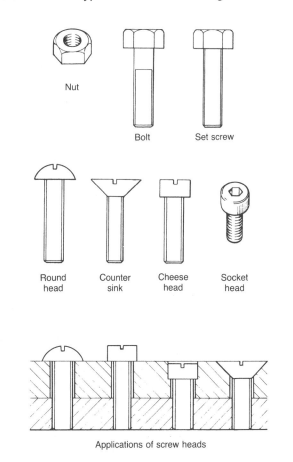

Nut

Bolt   Set screw

Round head   Counter sink   Cheese head   Socket head

Applications of screw heads

Figure 3.11 *Various types of nut, bolt and screw*

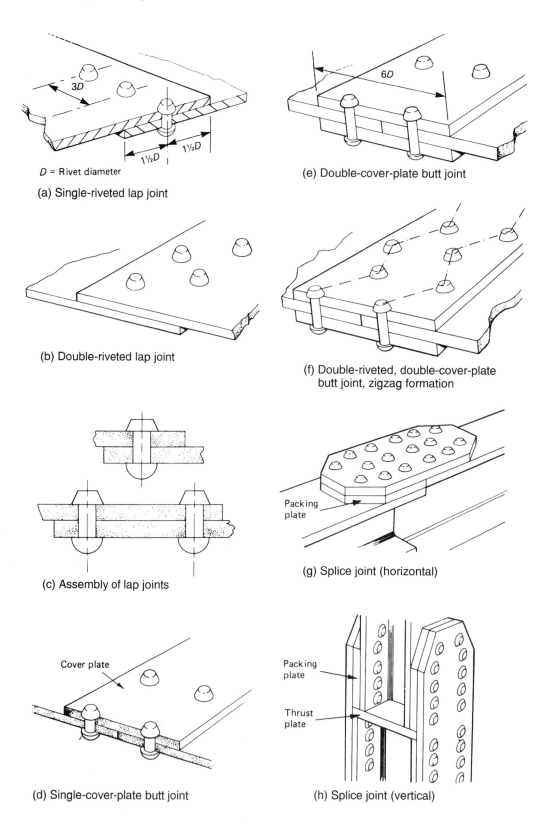

3D

D = Rivet diameter

1½D   1½D

(a) Single-riveted lap joint

(b) Double-riveted lap joint

(c) Assembly of lap joints

Cover plate

(d) Single-cover-plate butt joint

6D

(e) Double-cover-plate butt joint

(f) Double-riveted, double-cover-plate
butt joint, zigzag formation

Packing
plate

(g) Splice joint (horizontal)

Packing
plate

Thrust
plate

(h) Splice joint (vertical)

Figure 3.12 *Typical riveted joints*

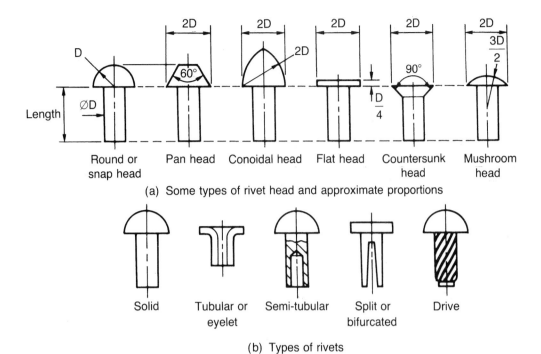

(a) Some types of rivet head and approximate proportions

(b) Types of rivets

Figure 3.13 *Typical rivet heads and types*

Electrical and electronic components

When selecting electrical and electronic components, you have to consider the following factors.

- Has the component got the correct circuit value? For example has it got the correct resistance or capacitance value. Also has it got the correct tolerance grade? The better the tolerance, the more expensive the component.
- Is it insulated to withstand the potential across the component and between the component and earth?
- If it is not enclosed, is it insulated against accidental contact (electric shock)?
- Can it pass the required current without overheating?
- Has it the correct power rating in watts?
- Will it fit onto the circuit board or chassis and is it suitable for connecting to the circuit board or associated components?
- If it is being used for telecommunications or data processing, is it suitable for the high frequencies used in these applications?

You have only to look through any electronics catalogue to see how many different types of electrical and electronic components are available. We only have room to look at a few items in this unit. Large scale manufacturers of electronic equipment would buy their components direct from the makers in bulk. For small scale batch production and for prototype work it is usual to buy from a wholesale or retail supplier. This enables you to obtain all your requirements on a 'one-stop' purchasing basis.

Figure 3.14 shows some typical cables and hardware for electronic equipment.

(a)   This shows examples of matrix board, strip board and a printed circuit board. The matrix board (1) is a panel of laminated plastic perforated with a grid of holes. Pins can be fixed in the holes at convenient places for the attachment of such components as resistors and capacitors. They are merely attachment points and do not form part of the circuit. The strip board (2) is like a matrix board but is copper faced in strips on one side. The holes are the same pitch as the pins of integrated circuits which can be soldered into position. The components are placed on the insulated side of the board, with the copper strips underneath. The wire leads pass through the holes in the board and are soldered on the underside. The copper tracks have to be cut wherever a break in the circuit is required. Printed circuit boards (3) are custom made for a particular circuit and are designed to give the most efficient layout for the circuit. The components are installed in the same way as for the strip board. Assembly will be considered in greater detail later in this unit.

(b)   This is a typical flexible mains lead with PVC sheathing and colour coded PVC insulation. In selecting such a cable, the only factors you have to consider is its current handling capacity and its colour, providing the insulation is rated for mains use.

(c)   This is a signal cable suitable for audio frequency analogue signals and for data processing signals. The conductors only need to have a limited current handling capacity, so many conductors can be carried in one cable. The conductors are surrounded by an earthed metal braid to prevent pick-up of external interference and corruption of the signal.

(d)   This is a ribbon cable widely used in the manufacture of computers and the interconnection of computers and printers. Unlike the signal cable described earlier it is not screened against interference. However, it is cheaper and more easily terminated.

(e)   Single cored PVC insulated wire is useful for making up wiring harnesses and for flying leads on PVC boards.

(f)   This shows a 'banana' plug and socket. These are used with single cored, flexible conductors for low-voltage power supply connections.

(g)   This shows a DIN type plug and socket. These are used in conjunction with multi-cored screened signal cables. They are available for 3-way to 8-way connections inclusive and are designed so that the plug can only be inserted into the socket in one position.

(h)   This show a 36-way Centronics plug as widely used for making connections to the parallel port of a computer printer.

(i)   This shows a 25-way D-type plug as widely used for making connections to the parallel output port of a computer.

(j)   This shows a selection of phono type plugs and sockets. These

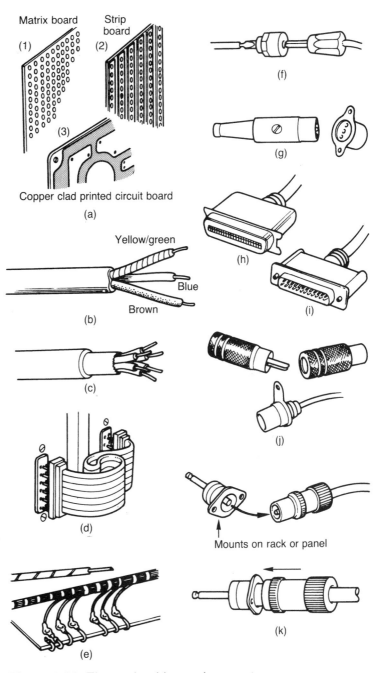

Figure 3.14 *Electronic wiring and connectors*

are widely used for making signal lead connections to audio amplifiers.

(k) This shows a typical plug and socket. These are used at radio frequencies for connecting aerial leads to television sets.

Figure 3.15 shows a selection of electronic components that are widely used. Resistors are used to limit the flow of an electric current in a circuit. Carbon type resistors can be used on both direct current and alternating current circuits at any frequency since they are noninductive. Wire wound resistors are inductive. They can only be used for direct current and mains frequency alternating current.

Capacitors are used to store electrical charges. Unlike batteries, they can be charged and discharged almost instantaneously. On the other hand, compared with batteries, they can only handle relatively small charges. Inductive devices such as chokes offer an impedance to alternating current. This increases as the frequency of the current gets higher. On direct current, inductive devices only offer the resistance of the wire from which they are wound.

(a) This shows a carbon rod resistor. They come in a wide range of resistance values and power ratings. These have quite wide tolerances but are suitable for general usage.

(b) This shows a high-stability resistor. These are precision resistors made from carbon film, metal film or oxide film. They are much more expensive than the ordinary carbon rod resistors. They are made to closer tolerances and are less susceptible to changes in resistance with changes in temperature. They are widely used in measuring instruments, computer applications and high stability radio frequency oscillators.

(c) This shows a wire-wound, vitreous enamelled resistor. These are inductive and only suitable for direct current, and mains frequency applications. They are used where high power ratings are required.

(d) This shows a typical carbon track variable resistor. These are used for volume control and tone control circuits.

(e) Carbon rod and high-stability resistors are too small for much information to be marked on them so they are usually coded in some way. The colour code is as follows:

				Tolerance	
0	Black	5	Green		
1	Brown	6	Blue	±5%	Gold
2	Red	7	Violet	±10%	Silver
3	Orange	8	Grey	±20%	No colour
4	Yellow	9	White	High stability	Pink

There are two ways of applying the colour code. Let's consider the old way first. This system had four colour bands. Three represented the resistance of the resistor and one represented the tolerance. Reading from the end of the resistor, the example in Figure 3.15(e) shows bands coloured red, violet, orange.

• The first band is the first number, red = 2
• The second band is the second number, violet = 7
• The third band is the number of zeros, orange = 3 or 000

So the resistance of our resistor is 27,000 Ω (or 27 kΩ)

If the resistor has a silver band as its fourth band, its resistance could range from 10% below its nominal value to 10% above its nominal value. That is, from 24,300 Ω to 29,700 Ω. Such a wide

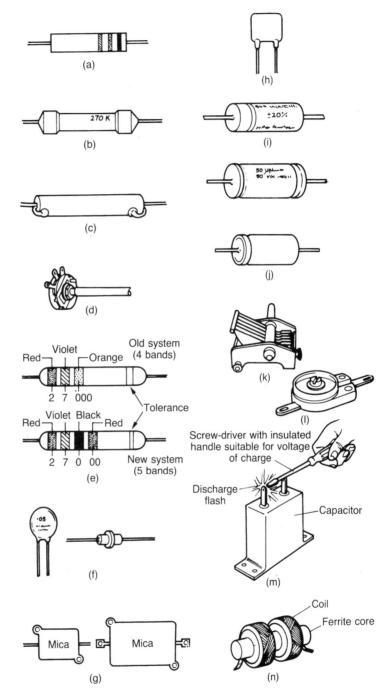

Figure 3.15 *A selection of typical electronic components*

range of values would be unacceptable for many applications, so additional close tolerance bands have been added; these are:

0.1%	Violet
0.25%	Blue
0.5%	Green
1%	Brown
2%	Red

The new system uses five bands, four for the resistance value and one for the tolerance. The example in Figure 3.15(e) shows bands of red, purple, black, red.

The first band is the first number, red = 2
The second band is the second number, violet = 7
The third band is the third number, black = 0
The fourth band is the number of zeros, red = 2 or 00

So the resistance of our resistor is once again 27,000 Ω.

If the resistor has a gold band as its fifth band, its resistance could range from 5% below its nominal value to 5% above its nominal value. That is, from 25,650 to 28,350 Ω. The additional band provides for intermediate values of resistance. For example if we had required 27,200 Ω ±1% the colours would have been red, violet, red, red, and a tolerance band coloured brown. This value could not have been achieved with the older system. Think about it.

A number and letter code is also used on circuit diagrams and often found printed on high stability resistors. This is best explained by some examples.

0.47 Ω	would be marked R47
4.7 Ω	would be marked 4R7
47 Ω	would be marked 47R
100 Ω	would be marked R100
1 kΩ	would be marked 1k0
10 kΩ	would be marked 10k
47 MΩ	would be marked 47M

Note that k = kilo = ×1,000 and that M = mega = ×1,000,000

(f) This shows examples of metallized ceramic capacitors. These are widely used in telecommunications equipment and in computers where high stability and compact size are required.

(g) This shows examples of silvered mica capacitors. These are also high stability capacitors suitable for radio frequency tuned circuits and for pulse operation.

(h) This shows a moulded polyester capacitor. These are self-healing and are widely used on printed circuit boards. They offer high values of capacitance in a small case size. They have a low inductance and low loss characteristics.

(i) This shows a polystyrene foil capacitor. These have ousted the foil and waxed paper capacitors found in some old equipment. They have low self-inductance, low high frequency losses and a long life. They are used for signal coupling and filter circuits.

(j) This shows some electrolytic capacitors. These are used where very high values of capacitance are required: for example smoothing capacitors in power packs. Normally these are polarized and they can only be used in direct current circuits. They must always be connected into the circuit in the correct direction as indicated on the case. Bi-polar electrolytic capacitors are available for use with low-voltage alternating currents and as signal coupling capacitors in audio amplifiers.

Write down the colour bands for each of the following resistance values using the four-band system.

(a)   56 Ω
(b)   470 Ω
(c)   33 kΩ
(d)   5.6 MΩ.

Write down the colour bands for each of the following resistance values using the five-band system.

(a)   562 Ω
(b)   675 kΩ.

A resistor has the following colour bands; orange, white, brown, gold. Determine its nominal value and its upper and lower limits of resistance.

Select suitable capacitor types for each of the following applications.

(a) A high stability radio frequency filter network
(b) a large value smoothing capacitor for a power supply unit
(c) a coupling capacitor for use in the signal amplifier circuit of a bass guitar.

(k) This shows a capacitor whose capacitance can be varied. Such a variable capacitor is wired in series or in parallel with an inductance (coil) to form a resonant (tuned) circuit. This is the way the tuning circuit of your radio works.

(l) This shows a preset pressure capacitor. The capacitance increases when the screw is rotated in a clockwise direction to tighten it. Such devices are used in pre-tuned circuits that do not have to be varied once they have been set.

(m) SAFETY: Large capacitors can store substantial charges of electricity at high voltages. Before handling such capacitors always discharge them as shown, either with a suitable screwdriver or with a length of insulated wire.

(n) This shows a typical inductor or coil. These may be air cored or they may be wound on a ferrite core to increase their inductance for a given size. Chokes have a single winding. Transformers may have a single winding with tappings (auto-transformer) or they may have two or more windings. They can only be used on alternating current circuits. Either power circuits or signal circuits.

Figure 3.16 shows some solid state devices.

(a) and (b) show some thermistors. The resistance of these devices falls off rapidly as the temperature increases. They can be used as sensors to activate thermal protection devices. A common application is as sensors in car engines to activate the temperature gauge on the instrument cluster. As the water temperature rises, the resistance of the sensor falls and the current through the circuit increases. This causes the temperature gauge to show a higher reading. Although measuring current, the scale of the instrument is calibrated in degrees of temperature.

(c) This shows some diodes. These are electronic switching devices that only allow the current to flow in one direction. To ensure that they are connected in the circuit the correct way round, the positive end is either chamfered, or it has a red band. On larger, metal-cased diodes the diode symbol is printed on the case with the arrow head pointing to the positive pole of the diode.

(d) This shows a selection of transistors. These are switching devices that allow a small current to control a larger current. Small changes in the applied current can cause corresponding and amplified changes in the larger current. Power transistors have metal cases that can be bolted to a heat-sink. If a transistor (or any other solid state device) overheats it is destroyed. For this reason heat-sinks should be clamped to the legs of solid state devices whilst they are being soldered into the circuit. The shape of the transistor case gives an indication as to the correct way to connect it into the circuit. Consult manufacturers' data sheets for the different shapes and the corresponding connections.

(e) This shows a typical integrated circuit. These contain many components on one chip. Complete amplifiers, radio receivers, timers and many other devices can come ready packaged in a single case ready for installation on a circuit board.

(a) **Rod type**
These special resistance elements have
a very high negative temperature
coefficient of resistance, making them
suitable as protective elements in a
wide range of circuits

Type	Resistance Cold	Resistance Hot	Dimensions in mm
TH-1A	650 Ω	37 Ω at 0.3 A	L. 38 Dia. 11
TH-2A	3.8 kΩ	44 Ω at 0.3 A	L. 32 Dia. 8
TH-3	370 Ω	28 Ω at 0.3 A	L. 22 Dia. 12
TH-5	4 Ω	0.4 Ω at 1 W (max)	Dia. 10 H. 4-5

(a)

(b) **Bead type**
Miniature glass-encapsulated
thermistors for amplitude control
and timing purposes (Types TH-B15,
TH-B18) or temperature measurement (Types TH-B11, TH-B12).
Selection tolerance ±20% at 20°C

Type	Resistance at 20 C	Minimum resistance	Dimensions in mm
TH-B11	1 MΩ	170 Ω	L10 Dia 2.5
TH-B12	2 kΩ	115 Ω	L10 Dia 2.5
TH-B15	100 kΩ	320 Ω	L25 Dia. 4
TH-B18	5 kΩ	100 Ω	L.38 Dia. 10

(b)

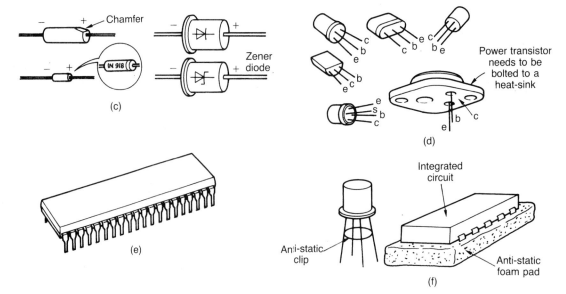

Figure 3.16 *A selection of thermistors and solid-state devices*

(f) I have already mentioned the care that must be taken to prevent a solid state device being destroyed by overheating. Equally important is the care that must be taken with some devices to prevent them being destroyed by electrostatic charges. Two ways of protecting such devices are shown in Figure 3.16(f). The anti-static clip must not be removed until the device has been installed in the circuit. Whilst installing such devices, the circuit board, the soldering iron and the installer must all be bonded to the same earth point. The installer wears a wrist clamp connected to earth via a flexible copper braid.

Health and safety

Health and safety legislation

The Health and Safety at Work, etc. Act of 1974 makes both the employer and the employee (you) equally responsible for safety. Both are equally liable to be prosecuted for violations of safety regulations and procedures. It is your legal responsibility to take reasonable care for your own health and safety. The law expects you to act in a responsible manner so as not to endanger yourself or other workers or the general public. It is an offence under the act to misuse or interfere with equipment provided for your health and safety or the health and safety of others.

Causes of accidents

Human carelessness

Most accidents are caused by human carelessness. This can range from 'couldn't care less' and 'macho' attitudes, to the deliberate disregard of safety regulations and codes of practice. Carelessness can also result from fatigue and ill-health resulting from a poor working environment.

Personal habits

Personal habits such as alcohol and drug abuse can render workers a hazard not only to themselves but also to other workers. Fatigue due to a second job (moonlighting) can also be a considerable hazard, particularly when operating machines. Smoking in prohibited areas where flammable substances are used and stored can cause fatal accidents involving explosions and fire.

Supervision and training

Another cause of accidents is lack of training or poor quality training. Lack of supervision can also lead to accidents if it leads to safety procedures being disregarded.

Environment

Unguarded and badly maintained plant and equipment are obvious causes of injury. However, the most common causes of accidents are falls on slippery floors, poorly maintained stairways, scaffolding and obstructed passageways in overcrowded workplaces. Noise, bad lighting, and inadequate ventilation can lead to fatigue, ill-health and carelessness. Dirty surroundings and inadequate toilet and washing facilities can lead to a lowering of personal hygiene standards.

Accident prevention

Elimination of hazards

The work place should be tidy with clearly defined passageways. It should be well lit and ventilated. It should have well maintained non-slip flooring. Noise should be kept down to acceptable levels. Hazardous processes should be replaced with less dangerous and more environmentally acceptable alternatives. For example asbestos clutch and brake linings should be replaced with safer materials.

Guards

Rotating machinery, drive belts and rotating cutters must be securely fenced to prevent accidental contact. Some machines have interlocked guards. These are guards coupled to the machine drive in such a way that the machine cannot be operated when the guard is open for loading and unloading the work. All guards must be set, checked and maintained by qualified and certified staff. They must not be removed or tampered with by operators. Some examples of guards are shown in Figure 3.17.

Maintenance

Machines and equipment must be regularly serviced and maintained by trained fitters. This not only reduces the chance of a major breakdown leading to loss of production, it lessens the chance of a major accident caused by a plant failure. Equally important is attention to such details as regularly checking the stocking and siting of first-aid cabinets and regularly checking the condition and siting of fire extinguishers. All these checks must be logged.

Personal protection

Suitable and unsuitable working clothing is shown in Figure 3.18. Some processes and working conditions demand even greater protection, such as safety helmets, earmuffs, respirators and eye protection worn singly or in combination. Such protective clothing must be provided by the employer when a process demands its use. Employees must, by law, make use of such equipment. Some examples are shown in Figure 3.19.

Safety education

This is important in producing positive attitudes towards safe working practices and habits. Warning notices and instructional posters should be displayed in prominent positions and in as many ethnic languages as necessary. Information, education and training should be provided in all aspects of health and safety: for example, process training, personal hygiene, first aid and fire procedures. Regular fire drills must be carried out to ensure that the premises can be evacuated quickly, safely and without panic.

Personal attitudes

It is important that everyone adopts a positive attitude towards safety. Not only your own safety but the safety of your workmates and the general public. Skylarking and throwing things about in the workplace or on site cannot be allowed. Any distraction that causes lack of concentration can lead to serious and even fatal accidents.

Housekeeping

A sign of a good worker is a clean and tidy working area. Only the minimum of tools for the job should be laid out at any one time. These tools should be laid out in a tidy and logical manner so that they immediately fall to hand. Tools not immediately required should be cleaned and properly stored away. All hand tools should be regularly checked and kept in good condition. Spillages, either on the workbench or on the floor should be cleaned up immediately.

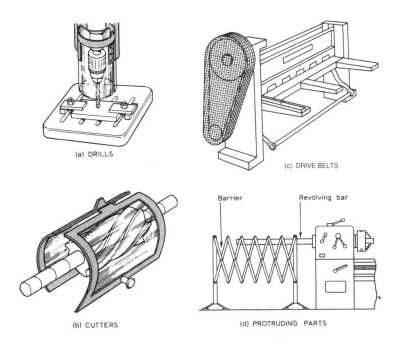

Figure 3.17 *Some examples of the use of guards*

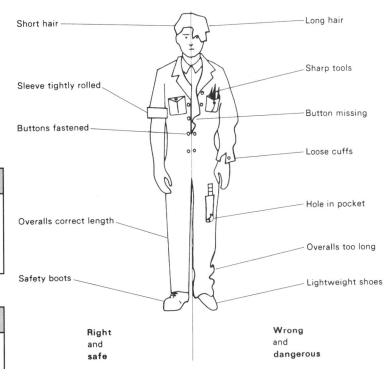

Figure 3.18 *Suitable and unsuitable clothing*

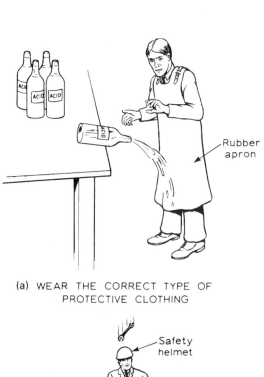

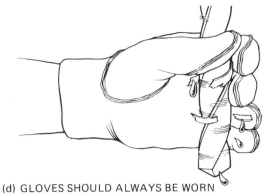

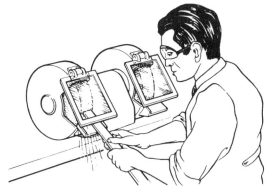

(d) GLOVES SHOULD ALWAYS BE WORN
WHEN HANDLING SHARP OBJECTS,
BUT NEVER WHEN OPERATING
MACHINE TOOLS

(a) WEAR THE CORRECT TYPE OF
PROTECTIVE CLOTHING

(b) PROTECT THE HEAD

(e) ALWAYS PROTECT THE EYES
WHEN USING MACHINERY

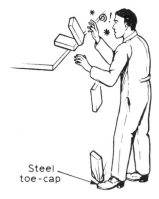

(c) WEAR SAFETY FOOTWEAR

(f) WEAR A SUITABLE RESPIRATOR
WHEN DUST AND FUMES ARE PRESENT

Figure 3.19 *Safety equipment and clothing*

Electrical hazards

Electrical equipment is potentially dangerous. The main hazards can be summarized as follows:

- electric shock
- fire due to the overheating of cables and equipment
- explosions set off by sparks when using unsuitable equipment when flammable vapours and gases are present.

Personal safety

Before using any electrical equipment it is advisable to carry out a number of visual checks as shown in Figure 3.20(a).

- Check that the cable is not damaged or frayed.
- Check that both ends of the cable are secured in the cord grips of the plug or appliance and that none of the conductors is visible.
- Check that the plug is in good condition and not cracked.
- Check that the voltage and power rating of the equipment is suitable for the supply available.
- If low voltage equipment is being used check that a suitable transformer is available.
- Check that whatever the voltage rating of the equipment, it is connected to the supply through a circuit-breaker containing a residual current detector (RCD).
- Check that all metal clad electrical equipment has a properly connected earth lead and is fitted with a properly connected three-pin plug as shown in Figure 3.20(b).

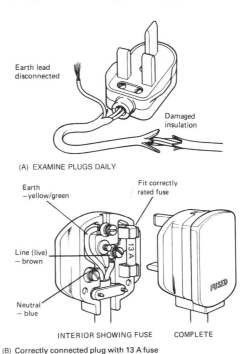

(A) EXAMINE PLUGS DAILY

Earth lead disconnected

Damaged insulation

Earth —yellow/green

Fit correctly rated fuse

Line (live) — brown

Neutral — blue

INTERIOR SHOWING FUSE   COMPLETE

(B) Correctly connected plug with 13 A fuse

Figure 3.20 *Visual checks on electric plugs and cables*

Test your knowledge 3.22

State THREE causes of electrical accidents and suggest how they may be prevented.

Test your knowledge 3.23

List the checks you would make before using a portable mains-operated electric power tool.

Earthing

All exposed metalwork of electrically powered or operated equipment must be earthed to prevent electric shocks (see also battery charger project). Figure 3.21 shows the two ways in which a person may receive an electric shock. In Figure 3.21(a) the person is receiving a shock by holding both the live and neutral conductors so that the electric current can flow through his or her body. The neutral conductor is connected to earth, so it is equally possible to receive a shock via an earth path when holding a live conductor, as shown in Figure 3.21(b). It is unlikely you would be so foolish as to deliberately touch a live conductor, but you might come into contact with one accidentally.

For example, the portable electric drill shown in Figure 3.22(a) has a metal casing but no earth. The live conductor has fractured within the machine and is touching the metal casing. The operator cannot see this and would receive a serious or even fatal shock. The fault current would flow through the body of the user via the earth path to neutral.

Figure 3.22(b) shows the effect of the same fault in a properly earthed machine. The fault current would take the path of least resistance and would flow to earth via the earth wire. The operator would be unharmed. Electrical equipment must be regularly inspected, tested, repaired and maintained by qualified electricians. Note that 'double insulated' equipment does not need to be earthed.

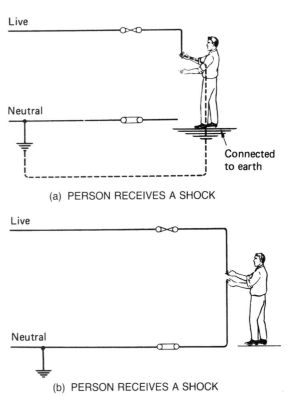

(a) PERSON RECEIVES A SHOCK

(b) PERSON RECEIVES A SHOCK

Figure 3.21 *Shock hazards*

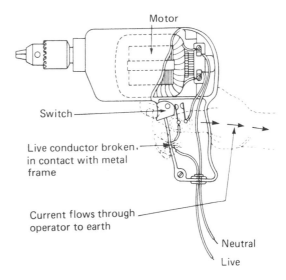

(a) ELECTRIC DRILL NOT EARTHED

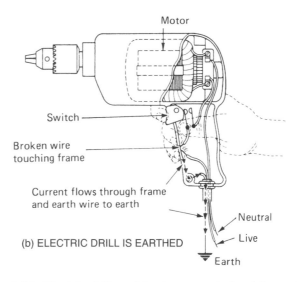

(b) ELECTRIC DRILL IS EARTHED

Figure 3.22 *Electric drill earthing to prevent shock hazard*

Procedure in the event of electric shock

- Switch off the supply of current if this can be done quickly.
- If you cannot switch off the supply, do not touch the person's body with your bare hands. Human flesh is a conductor and you would also receive a severe shock. Drag the affected person clear using insulating material such as dry clothing, a dry sack, or any plastic material that may be handy.
- If the affected person has stopped breathing commence artificial respiration immediately. Don't wait for help to come or go to seek for help. If the affected person's pulse has stopped, heart massage will also be required. Use whatever method of resuscitation with which you are familiar.

Fire

Fire is the rapid oxidation (burning) of flammable materials. For a fire to start, the following are required:

- a supply of flammable materials
- a supply of air (oxygen)
- a heat source.

Once the fire has started, the removal of one or more of the above will result in the fire going out.

Fire prevention

Fire prevention is largely 'good housekeeping'. The workplace should be kept clean and tidy. Rubbish should not be allowed to accumulate in passages and disused storerooms. Oily rags and waste materials should be put in metal bins fitted with airtight lids. Plant, machinery and heating equipment should be regularly inspected, as should fire alarm and smoke detector systems. You should know how to give the alarm.

Electrical installations, alterations and repairs must only be carried out by qualified electricians and must comply with the current IEE Regulations. Smoking must be banned wherever flammable substances are used or stored. The advice of the fire prevention officer of the local brigade should be sought before flammable substances, bottled gases, cylinders of compressed gases, solvents and other flammable substances are brought on site.

Fire procedures

In the event of you discovering a fire, you should:

- Raise the alarm and call the fire service.
- Evacuate the premises. Regular fire drills must be held. Personnel must be familiar with normal and alternative escape routes. There must be assembly points and a roll call of personnel. A designated person must be allocated to each department or floor to ensure that evacuation is complete. There must be a central reporting point.
- Keep fire doors closed to prevent the spread of smoke. Smoke is the biggest cause of panic and accidents, particularly on staircases. Emergency exits must be kept unlocked and free from obstruction whenever the premises are in use. Lifts must not be used in the event of fire.
- Only attempt to contain the fire until the professional brigade arrives if there is no danger to yourself or others. Always make sure you have an unrestricted means of escape. Saving lives is more important than saving property.

Extinguishers

Figure 3.23 shows a fire hose and a range of pressurized water extinguishers. These can be identified by their shape and colour which is RED. They are for use on burning solids such as wood, paper, cloth, etc. They are UNSAFE on electrical equipment at all voltages.

Figure 3.24 shows two types of foam extinguisher. These can be identified by their shape and colour which is CREAM. They are for use on burning flammable liquids. They are UNSAFE on electrical equipment at all voltages.

Figure 3.25 shows a variety of extinguishers that can be used on most fires and are safe for use on electrical equipment. Again, they can be identified by their shapes and colours:

- Dry powder extinguishers are coloured BLUE and are safe up to 1,000 V.
- Carbon dioxide (CO_2) extinguishers are coloured BLACK and are safe at high voltages.
- Vaporizing liquid extinguishers are coloured GREEN and are safe at high voltages.

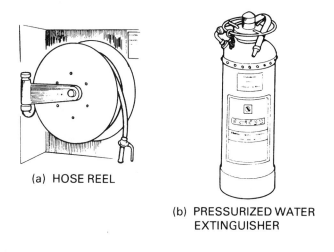

(a) HOSE REEL

(b) PRESSURIZED WATER EXTINGUISHER

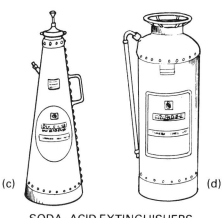

(c)   (d)

SODA–ACID EXTINGUISHERS

Figure 3.23 *Various types of fire extinguisher*

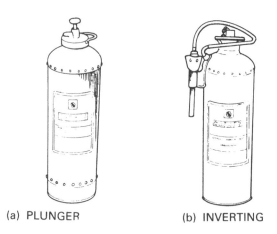

(a) PLUNGER   (b) INVERTING

Figure 3.24 *Foam extinguishers*

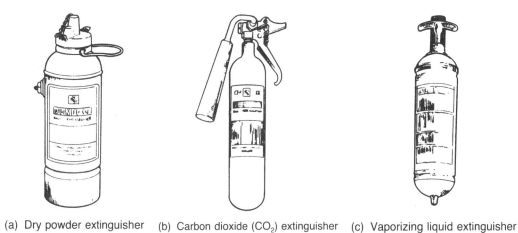

(a) Dry powder extinguisher   (b) Carbon dioxide (CO_2) extinguisher   (c) Vaporizing liquid extinguisher

Figure 3.25 *Extinguishers for use on electrical equipment*

These latter two extinguishers act by replacing the air with an atmosphere free from oxygen. They are no good in draughts which would blow the vapour or gas away. Remember, if the fire can't breathe neither can any living creature. Evacuate all living creatures before using one of these types of extinguisher. When using this type of extinguisher, keep backing away from the gas towards fresh air, otherwise it will put you out as well as the fire!

Figure 3.26 shows a fire blanket. Fire blankets are woven from fire-resistant synthetic fibres and are used to smother fires. The old-fashioned blankets made from asbestos must NOT be used. The blanket is pulled from its container and spread over the fire to exclude the air necessary to keep the fire burning. They are suitable for use in kitchens, in workshops, and in laboratories. They are also used where a person's clothing is on fire, by rolling the person and the burning clothing up in the blanket to smother the fire. Do not cover the person's face.

Test your knowledge 3.24

State which type of fire extinguisher you would use in each of the following cases:

(a) Paper burning in an office waste bin
(b) A pan of fat burning in the kitchen of the works canteen
(c) A fire in a mains voltage electrical machine.

Figure 3.26 *A fire blanket*

Test your knowledge 3.25

A fire breaks out near to a store for paints, paint thinners and bottled gases. What action should be taken and in what order?

Activity 3.5

Consult the Health and Safety at Work Act and answer the following questions:

(a) What is an improvement notice?
(b) What is a prohibition notice?
(c) Who issues such notices?
(d) Who can be prosecuted under the Act?

Present your findings in the form of a brief word processed report.

Activity 3.6

Investigate the construction of one of the lathes in your workshop and describe how the guard over the end train gears (change wheels) is interlocked so that the lathe cannot operate with the guard open. Illustrate your findings with a sketch.

Activity 3.7

Design a poster that can be placed in a workshop that will provide information on the types of fire extinguisher that can be safely used on different types of fire. Your poster should be designed so that it can be understood by people with a limited command of the English language.

Measuring

Before you can mark out a component or check it during manufacture you need to know about engineering measurement. All engineering measurements are comparative processes. You compare the size of the feature to be measured with a known standard.

Linear measurement

Figure 3.27 shows a steel rule and how to use it. The distance between the lines or the width of the work is being compared with the rule. In this instance the rule is our standard of length. The rule should be made from spring steel and the markings should be engraved into the surface of the rule. The edges of the rule should be

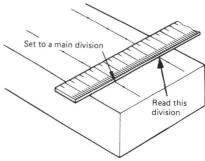

(a) MEASURING THE DISTANCE BETWEEN TWO SCRIBED LINES

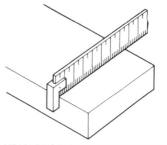

(b) MEASURING THE DISTANCE BETWEEN TWO FACES USING A HOOK RULE

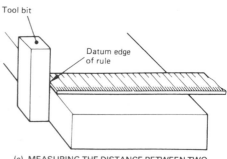

(c) MEASURING THE DISTANCE BETWEEN TWO FACES USING A STEEL RULE AND ABUTMENT

Figure 3.27 *Using a steel rule*

ground so that it can be used as a straight edge, and the datum end of the rule should be protected from damage so that the accuracy of the rule is not lost. NEVER use a rule as a screwdriver or for cleaning out the T-slots on machine tools.

To increase the usefulness of the steel rule and to improve the accuracy of taking measurements, accessories called calipers are used. These are used to transfer the distances between the faces of the work to the distances between the lines engraved on the rule. Figure 3.28 shows some different types of inside and outside calipers. It also shows how to use calipers.

A steel rule can only be read to an accuracy of about ± 0.5 mm. This is rarely accurate enough for precision engineering purposes. Figure 3.29(a) shows a vernier caliper and how it can take inside and outside measurements. Typical vernier scales are shown in Figure 3.29(b). Some verniers have scales that are different to the ones shown in Figure 3.29(b). Always check the scales before taking a reading. Like all measuring instruments, a vernier caliper must be treated carefully and it must be cleaned and returned to its case whenever it is not in use.

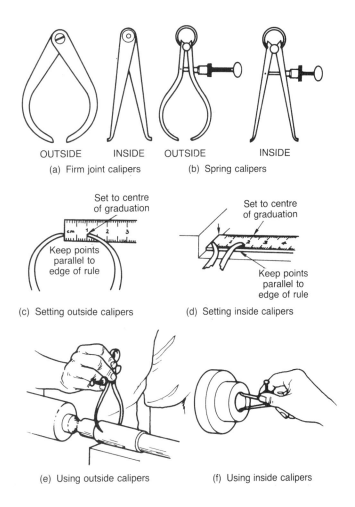

Figure 3.28 *Inside and outside calipers*

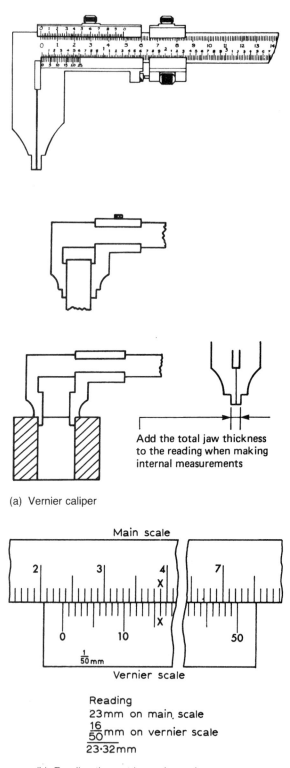

(a) Vernier caliper

Main scale

Vernier scale

Reading
23mm on main. scale
$\frac{16}{50}$mm on vernier scale
23·32mm

(b) Reading the metric vernier scale:
    23 mm on main scale plus 16 × 0.02 mm
    on vernier scale gives total 23.32 mm

Figure 3.29 *Vernier calipers*

Vernier calipers are difficult to read accurately even if you have good eyesight. A magnifying glass is helpful. The larger sizes of vernier caliper are quite heavy and it is difficult to get a correct and consistent 'feel' between the instrument and the work. An alternative instrument is the micrometer caliper. Figure 3.30 shows a micrometer caliper and the method of reading it's scales.

Micrometer calipers are more compact than vernier calipers. They are also easier to use. The ratchet on the end of the thimble ensures that the contact pressure is kept constant and at the correct value. Unfortunately micrometer calipers have only a limited measuring range (25 mm), so you need a range of micrometers moving up in size in 25 mm steps. (0—25 mm, 25—50 mm, 50—75 mm and so on to the largest size). You also need a range of inside micrometers, and depth micrometers.

Angular measurements

The most frequently measured angle is 90°. This is a right-angle and surfaces at right-angles to each other are said to be perpendicular. Right angles are checked with some sort of try square. Figure 3.32 shows a typical engineer's try square and two ways in which it can be used.

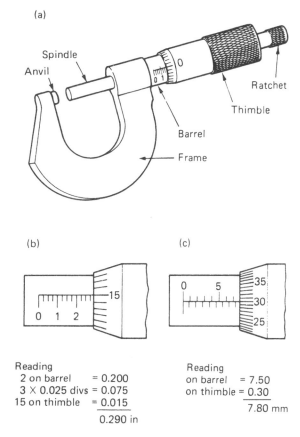

(a)

Spindle

Anvil

Ratchet

Thimble

Barrel

Frame

(b)

(c)

Reading
2 on barrel   = 0.200
3 × 0.025 divs = 0.075
15 on thimble   = 0.015
                        0.290 in

Reading
on barrel   = 7.50
on thimble = 0.30
                   7.80 mm

Figure 3.30 *Micrometer caliper*

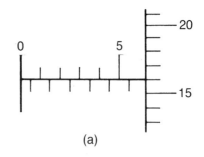

(a)

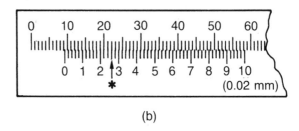

(b)

Figure 3.31 *See Test your knowledge 3.28*

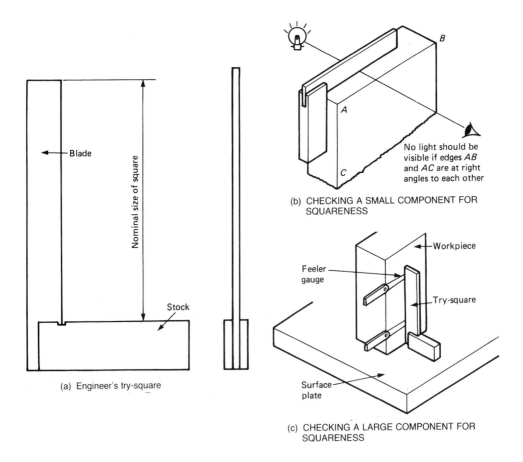

(a) Engineer's try-square

(b) CHECKING A SMALL COMPONENT FOR SQUARENESS

No light should be visible if edges *AB* and *AC* are at right angles to each other

(c) CHECKING A LARGE COMPONENT FOR SQUARENESS

Figure 3.32 *Engineer's try square*

Test your knowledge 3.29

Name the instrument that is used to check that two surfaces are at right-angles to one another.

Test your knowledge 3.30

State the reading accuracy of the vernier bevel protractor shown in Figure 3.33(b).

Test your knowledge 3.31

Explain how the reading stated in Figure 3.33(b) is obtained.

For angles other than a right-angle, a protractor is used. This may be a simple plain protractor as shown in Figure 3.33(a) or it may have a vernier scale (vernier protractor) as shown in Figure 3.33(b).

Tolerance and gauging

So far we have only considered measurement of size. This is usual when making a small number of components. However for quantity production it requires too high a skill level, is too time consuming and, therefore, too expensive. Since no product can be made to an exact size, nor can it be measured exactly, the designer usually gives each dimension an upper and lower size. This is shown in Figure 3.34. If the component lies anywhere between the upper and lower

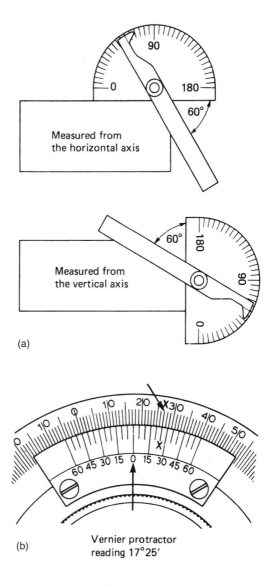

Figure 3.33 *Using a protractor*

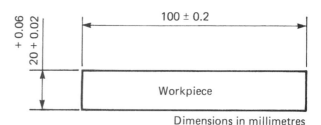

Dimensions in millimetres

EXAMPLE 1

Nominal size	100 mm
Limits (low)	99.8 mm
Limits (high)	100.2 mm
Tolerance	0.4 mm
Deviation	± 0.2 mm
Mean size	100.0 mm

EXAMPLE 2

Nominal size	20 mm
Limits (low)	20.2 mm
Limits (high)	20.6 mm
Tolerance	0.4 mm
Deviation	+0.02, +0.06
Mean size	20.4 mm

Figure 3.34 *Use of tolerances*

Test your knowledge 3.32

With reference to Figure 3.37, state the following:

(a) the nominal size
(b) the upper limit
(c) the lower limit
(d) the tolerance
(e) the deviation
(f) the mean size.

Test your knowledge 3.33

With the aid of sketches, explain how you would check the dimensions for the hole shown in Figure 3.37 with a plug gauge.

limits of size it will function correctly. The closer the limits, the more accurately the component will work, but the more expensive it will be to make.

A major advantage of using tolerance dimensions is that they can be checked without having to be measured. Gauges can be used instead of measuring instruments. This is easier, quicker and much cheaper. Figure 3.35 shows how a caliper gauge can be used to check the thickness of a component. Plug gauges are used in a similar manner to check hole sizes.

Some more gauges are shown in Figure 3.36. Radius gauges are used to check the corner radii of components. Feeler gauges are used to check the gap between components: for example the valve tappet clearances in a motor vehicle engine. Thread gauges are used to check the pitch of screw threads.

Activity 3.8

Select suitable measuring instruments and explain how you would check the dimensions for the component shown in Figure 3.38.

Write down the readings for:
(a) the micrometer shown in Figure 3.39(a)
(b) the vernier shown in Figure 3.39(b).

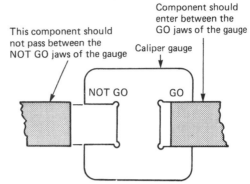

(a) CORRECTLY SIZED COMPONENT ENTERS 'GO'
JAWS BUT NOT 'NOT GO' JAWS

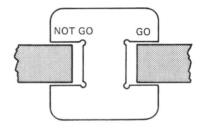

(b) UNDERSIZE COMPONENT ENTERS 'GO' AND
'NOT GO'

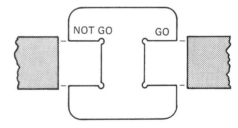

(c) OVERSIZE COMPONENT DOES NOT ENTER 'GO'
OR 'NOT GO'

Figure 3.35 *Using a caliper gauge to check the thickness of a component*

Marking out

The reasons why you mark out components before making them are as follows.

- To provide you with guide lines to work to. Where only limited accuracy is required, marking out also controls the size and shape of the workpiece and the position of any holes.
- As a guide to a machinist when setting up and cutting. In this instance the final dimensional control comes from the use of precision measuring instruments together with the micrometer dials on the machine controls.
- Marking out also indicates if sufficient machining allowance has been left on cast or forged components. It also indicates whether or not such features as webs, flanges and cored holes have been correctly positioned.

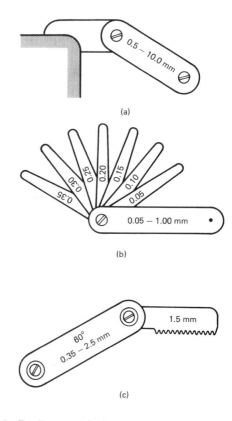

Figure 3.36 *Radius and feeler gauges*

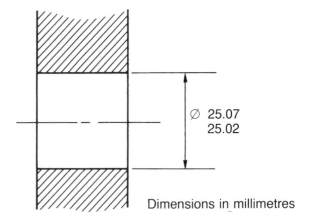

Figure 3.37 *See Test your knowledge 3.32 and 3.33*

Scribed lines and centre marks

Scribed lines are fine lines cut into the surface of the material being marked out by the point of a scribing tool (scriber). An example of a scriber is shown in Figure 3.40(a). To ensure that the scribed line shows up clearly, the surface of the material to be marked out is coated with a thin film of a contrasting colour. For example, the

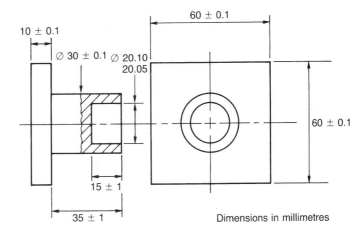

Figure 3.38 *See Activity 3.8*

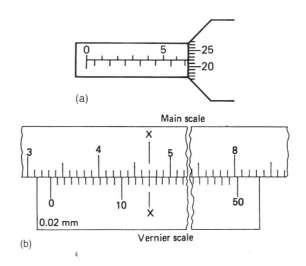

Figure 3.39 *See Activity 3.8*

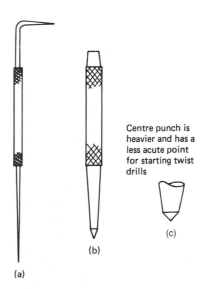

Centre punch is heavier and has a less acute point for starting twist drills

Figure 3.40 *Scriber and centre punch*

surfaces of the casting to be machined are often whitewashed. Bright metal surfaces can be treated with a marking out 'ink'. Plain carbon steels can be treated with copper sulphate solution which copper plates the surface of the metal. This has the advantages of permanence. Care must be taken in its use as it will attack any marking out and measuring instruments into which the copper sulphate comes into contact.

Centre marks are made with a dot punch as shown in Figure 3.40 (b) or with a centre punch, as shown in Figure 3.40(c). A dot punch has a fine conical point with an included angle of about 60°. A centre punch is heavier and has a less acute point angle of about 90°. It is used for making a centre mark for locating the point of a twist drill and preventing the point from wandering at the start of a cut.

The dot punch is used for two purposes when marking out.

- A scribed line can be protected by a series of centre marks made along the line, as shown in Figure 3.41(a). If the line is accidentally removed, it can be replaced by joining up the centre marks. Further, when machining to a line as shown in Figure 3.41 (b), the half-marks left behind are a witness that the machinist has 'split the line'.
- Secondly, dot punch marks are used to prevent the centre point of dividers from slipping when scribing circles and arcs of circles, as shown in Figure 3.41(c). The correct way to set divider points is shown in Figure 3.41(d).

When a centre punch is driven into the work, distortion can occur. This can be a burr raised around the punch mark, swelling of the edge of a component, or the buckling of thin material.

Equipment for marking out

From what we have already seen, your basic requirements for marking out are:

- A scriber to produce a line
- A rule to measure distances and act as a straight edge to guide the point of the scriber
- Dividers to scribe circles and arcs of circles as shown in Figure 3.41(c).

In addition, you require hermaphrodite (odd-leg) calipers, as shown in Figure 3.42(a) and a try square. Odd-leg calipers are used to scribe lines parallel to a datum edge. A try square and scriber are used to scribe a line at right-angles to a datum edge, as shown in Figure 3.42 (b).

Alternatively, lines can be scribed parallel to a datum edge using a surface table and scribing block, as shown in Figure 3.44(a). In this example the scribing point is set to a steel rule, so the accuracy is limited. Alternatively, the line can be scribed using a vernier height gauge, as shown in Figure 3.44(b). This is very much more accurate. Where extreme accuracy is required, slip gauges and slip gauge accessories can be used, as shown in Figure 3.44(c).

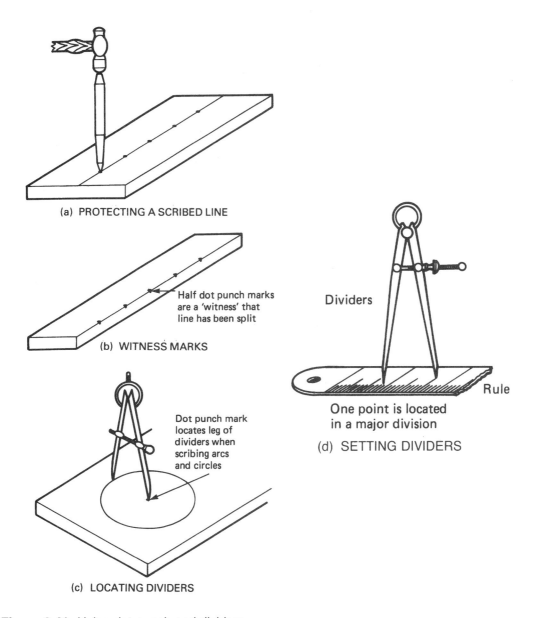

(a) PROTECTING A SCRIBED LINE

Half dot punch marks are a 'witness' that line has been split

(b) WITNESS MARKS

Dot punch mark locates leg of dividers when scribing arcs and circles

(c) LOCATING DIVIDERS

Dividers

Rule

One point is located in a major division

(d) SETTING DIVIDERS

Figure 3.41 *Using dot punch and dividers*

Cylindrical components are difficult to mark out since they tend to roll about. To prevent this they can be supported on vee-blocks, as shown in Figure 3.45(a). Vee-blocks are always made and sold in boxed sets of two. In order that the axis of the work is parallel to the surface plate or table, you must always use such a matched pair of vee-blocks, and make sure that, after use, they are always put away as a pair.

Another useful device for use on cylindrical work is the box square shown in Figure 3.45(b). This is used for scribing lines along cylindrical work parallel to the axis. A centre finder is shown in Figure 3.45(c). This is used to scribe lines that pass through the centre of circular blanks or the ends of cylindrical components.

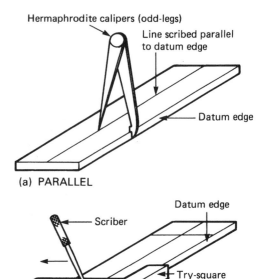

Figure 3.42 *Using odd-leg calipers, scriber and try square*

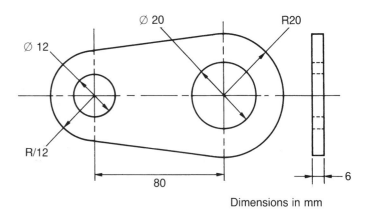

Figure 3.43 *See Test your knowledge 3.34*

Datum points, lines and surfaces

First we had better revise what we know about rectangular and polar coordinates. Examples of these are shown in Figure 3.46.

Rectangular coordinates

The point A in Figure 3.46(a) is positioned by a pair of ordinates (coordinates) lying at right-angles to each other. They also lie at right-angles to the datum edges from which they are measured. This system of measurement requires the production of two datum surfaces or edges at right-angles to each other. That is, two datum edges that are mutually perpendicular.

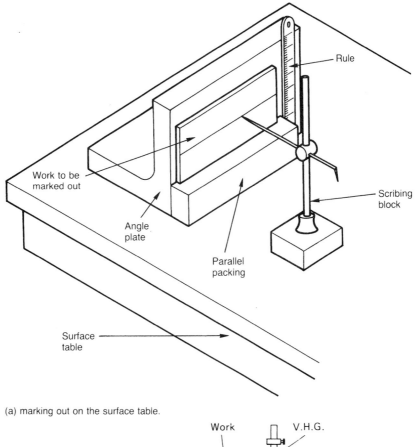

(a) marking out on the surface table.

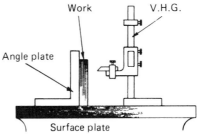

(b) Marking out with the vernier height

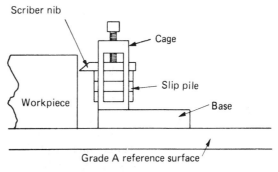

(c) Marking out using slip gauges and accessories

Figure 3.44 *Marking out using a surface table*

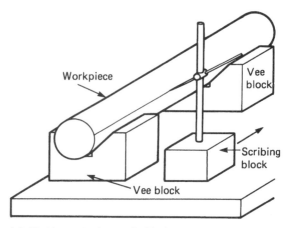

(a) Marking out a long cylindrical component on vee blocks

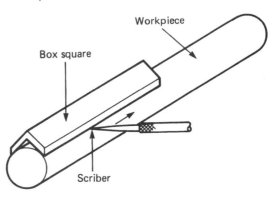

(b) Use of the box square

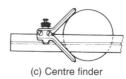

(c) Centre finder

Figure 3.45 *Marking cylindrical components*

Polar coordinates

Polar coordinates consist of one linear distance and an angle, as shown in Figure 1.46(b). Dimensioning in this way is useful when the work is to be machined on a rotary table. It is widely used when setting out hole centres round a pitch circle. Quite frequently both systems are used at the same time, as shown in Figure 3.46(c)

During our discussion on marking out, I have kept referring to datum points, datum lines and datum edges. A datum is any point, line or surface that can be used as a basis for measurement. When you go for a medical check-up, your height is measured from the floor on which you are standing. In this example the floor is the basis of measurement, it is the datum surface from which your height is measured.

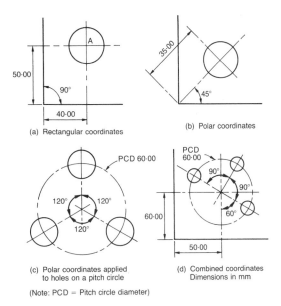

(a) Rectangular coordinates

(b) Polar coordinates

(c) Polar coordinates applied to holes on a pitch circle

(d) Combined coordinates Dimensions in mm

(Note: PCD = Pitch circle diameter)

Figure 3.46 *Datum points and coordinates*

Point datum

This is a single point from which a number of features are marked out. For example, Figure 3.47(a) shows two concentric circles representing the inside and ouside diameter of a pipe-flange ring. It also shows the pitch circle around which the bolt hole centres are marked off. All these are marked out using dividers or trammels (beam compasses) from a single-point datum.

Line datum

Any line from which, or along which a number of features are marked out. An example of the use of line datums is shown in Figure 3.47(b).

Surface datum

This is also known as an edge datum and a service edge. It is the most widely used datum for marking out solid objects. Two edges are accurately machined at right angles to each other and all the dimensions are taken from these edges. Figure 3.47(c) shows how dimensions are taken from surface datums. It shows both rectangular and polar coordinates. Alternatively, the work can be clamped to an angle-plate. The mutually perpendicular edges of the angle-plate provide the surface datums. In this instance there is no need to machine the edges of the workpiece at right-angles.

Activity 3.9

With the aid of sketches explain how you would mark out the holes and the square on the component shown in Figure 3.48. Present your answer in the form of an instruction sheet that is suitable for use by someone with no previous experience of this type of task.

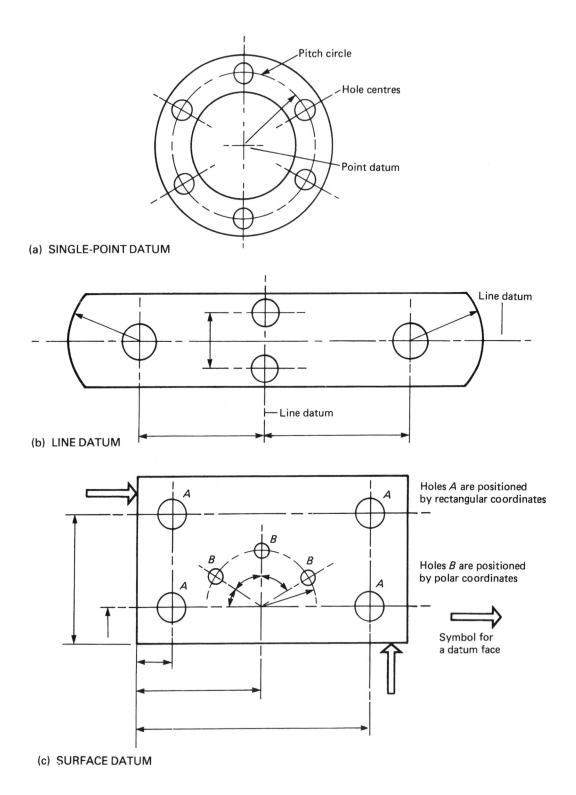

Figure 3.47 *Point, line and surface datums*

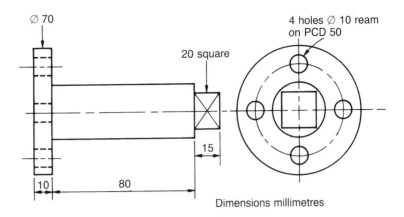

Dimensions millimetres

Figure 3.48 *See Activity 3.9*

Hand tools and benchwork

The next section introduces you to a variety of hand tools and how they are used. We will start with the fitter's bench and vice that is used for holding work whilst performing cutting, chiseling or filing.

Fitter's bench and vice

A fitter's bench should be substantial and rigid. This is essential if accurate work is to be performed on it. It should be positioned so that it is well lit by both natural and artificial light without glare or shadows. It should be equipped with a fitter's vice. A plain screw vice is shown in Figure 3.49(a) and a quick action vice is shown in Figure 3.49(b). In the latter type of vice, the jaws can be quickly pulled apart when the lever at the side of the screw handle is

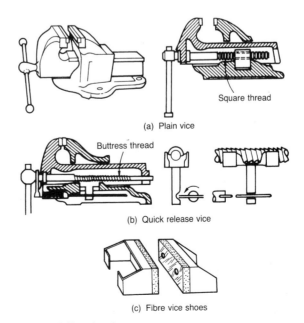

(a) Plain vice

(b) Quick release vice

(c) Fibre vice shoes

Figure 3.49 *A fitter's vice*

released. The screw is used for closing the jaws and clamping the work in the usual way.

The jaws of a fitter's vice are serrated and hardened to prevent the work from slipping. This also marks the surfaces of the work. For fine work with finished surfaces, the serrated jaws should be replaced with hardened and ground smooth jaws. Alternatively vice shoes can be used. These are faced with a fibre compound and can be slipped over the serrated jaws when required. A pair of typical vice shoes are shown in Figure 3.49(c).

Cutting tools

Before we can discuss the cutting tools we use for bench fitting we need to look at the way metal is cut. Here are the basic facts.

Wedge angle
If you look at a hacksaw blade, as shown in Figure 3.50(a), you can see that the teeth are wedge shaped. Figure 3.50(b) shows how the wedge angle increases as the material gets harder. This strengthens the cutting edge and increases the life of the tool. At the same time it reduces its ability to cut. Try cutting a slice of bread with a cold chisel!

Clearance angle
If you look at the hacksaw blade in Figure 3.50(a), you can see that there is a clearance angle behind the cutting edge of the tooth. This is to enable the tooth to cut into the work.

Rake angle
This angle controls the cutting action of the tool. It is shown in Figure 3.51. I hope you can see that the wedge angle, clearance angle and rake angle always add up to 90°. This is true even when the rake angle is zero or negative as shown in Figure 3.51(b). The greater the rake angle the more easily the tool will cut. Unfortunately, the greater the rake angle, the smaller the wedge

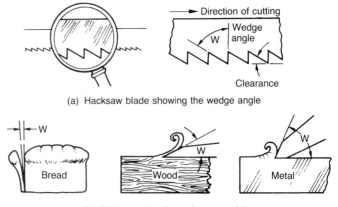

(a) Hacksaw blade showing the wedge angle

(b) Wedge angles for various materials

Figure 3.50 *Hacksaw blade and wedge angles*

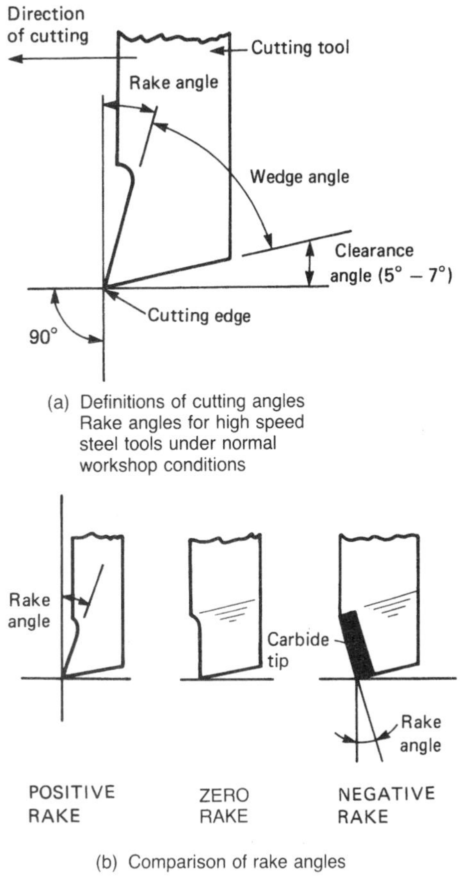

(a) Definitions of cutting angles
    Rake angles for high speed
    steel tools under normal
    workshop conditions

(b) Comparison of rake angles

Figure 3.51 *Rake angle*

angle will be and the weaker the tool will be. Therefore the wedge and rake angles have to be a compromise between ease of cutting and tool strength and life. The clearance angle remains constant at between 5° and 7°.

Table 3.7 *Rake angle for high speed tools under normal conditions*

Material	Rake angle
Aluminium alloy	30°
Brass (ductile)	14°
Brass (free-cutting)	0°
Cast iron	0°
Copper	20°
Phosphor bronze	8°
Mild steel	25°
Medium carbon steel	15°

Orthogonal and oblique cutting

Figure 3.52(a) is a pictorial representation of the single point cutting tool shown in Figure 3.51. Notice how the cutting edge is at right-angles to the direction in which the tool is travelling along the work. This is called orthogonal cutting. Now look at Figure 3.52(b). Notice how the cutting edge is inclined at an angle to the direction of cut. This is called oblique cutting. Oblique cutting results in a better finish than orthogonal cutting, mainly because the chip is thinner for a given rate of metal removal. This reduced thickness and the geometry of the tool allows the chip to coil up easily in a spiral.

Apart from threading operations, it is very rare to use a coolant or lubricant when using hand tools. However the conditions are very different when using machine tools. Large amounts of metal are removed quickly, considerable energy is used to do this, and this energy is largely converted into heat at the cutting zone. The rapid temperature rise of the work and the cutting tool can lead to inaccuracy and short tool life. A coolant is required to prevent this. The chip flowing over the rake face of the tool results in wear. A lubricant is required to prevent this. Usually coolants are poor lubricants, and lubricants are poor coolants.

For general machining an emulsion of cutting oil (which also contains an emulsifier) and water is used. This has a milky-white appearance and is commonly known as 'suds'. On no account try to use a mineral lubricating oil. This cannot stand up to the temperatures and pressures found in the cutting zone. It is completely useless as a cutting lubricant or as a coolant. It gives off clouds of noxious fumes and it is a fire risk.

Cold chisels

Let's now see how the cutting angles discussed in the previous section can be applied to the basic bench tools.

Figure 3.53(a) shows a typical cold chisel and names its more important features. Constantly hitting the head of a chisel causes it to mushroom, as shown in Figure 3.53(b). Never use a chisel with a mushroom head because bits of metal can fly off it when it is hit.

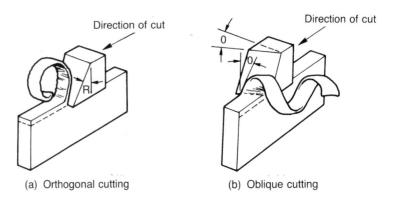

(a) Orthogonal cutting   (b) Oblique cutting

Figure 3.52 *Orthogonal and oblique cutting*

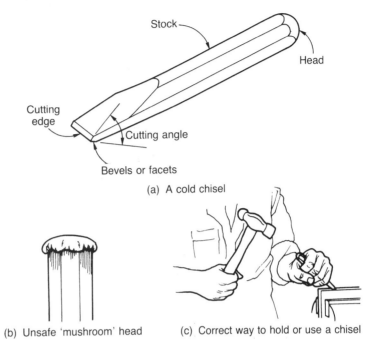

(a) A cold chisel

(b) Unsafe 'mushroom' head   (c) Correct way to hold or use a chisel

Figure 3.53 *Using a cold chisel*

These can cause an accident. When the mushroom head starts to form, it must be trimmed off on a grinding machine. Figure 3.53(c) shows the correct way to hold and use a cold chisel.

Safety when chipping

There are several important safety points that should always be observed when using a cold chisel:

- Never chip towards another person
- Always chip towards a chipping screen
- Always wear goggles when chipping
- Always grind the mushroom head off a chisel before using it and make sure the cutting edge is sharp and in good condition. Regrind if necessary.

The chisel shown in Figure 3.53 is only one of many different types of chisel. Some further examples and their applications are shown in Figure 3.54. In the course of a working lifetime most fitters will make up many small chisels for special jobs. These are often made from hardened and tempered silver steel rod. In addition there are the fine engraving chisels or 'gravers' used by die-sinkers and engravers.

The application of the cutting angles we discussed earlier can be applied to a chisel as shown in Figure 3.55. The point angle (wedge angle) and the angle of inclination are only a guide. In practice, a fitter does not work out the angle of inclination or the rake and clearance angle, but uses experience and the feel of the chisel as it cuts through the metal to present the chisel to the work at the correct angle.

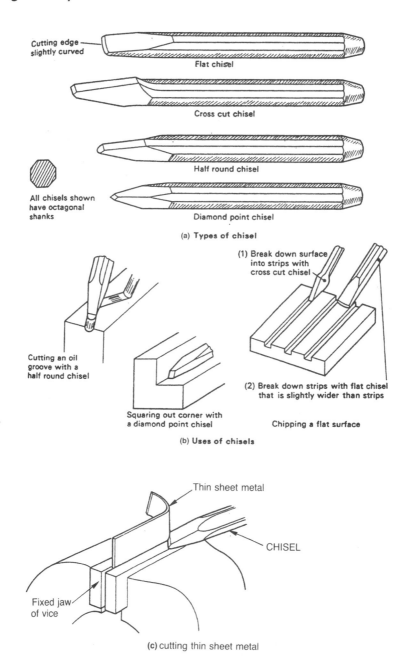

Figure 3.54 *Types of chisel and applications*

Files and filing

Files are the most widely used and important tools for the fitter. The main parts of a file are named in Figure 3.56(a). Files are forged to shape from 1.2% plain carbon steel. After forging, the teeth are machine cut by a chisel shaped tool, as shown in Figure 3.56(b). The teeth of a single-cut are wedge shaped with the rake and clearance angles essential for metal cutting. Most files used in general engineering are double-cut. That is they have two rows of cuts at an angle to each other, as shown in Figure 3.56(c).

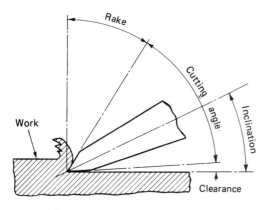

(a) CUTTING ACTION OF CHISEL

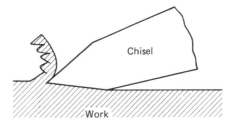

(b) ANGLE OF INCLINATION TOO SMALL
CHISEL POINT RISES

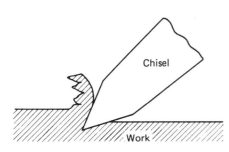

(c) ANGLE OF INCLINATION TOO GREAT
CHISEL POINT DIGS IN

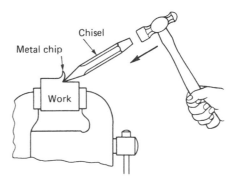

(d) CUT TOWARDS FIXED JAW OF VICE

Typical cutting and inclination angles (clearance angle constant at 7°):

Material	Point Angle	Angle of inclination
Cast iron	60°	37°
Mild steel	55°	34.5°
Medium carbon steel	65°	39.5°
Brass	50°	32°
Copper	45°	29.5°
Aluminium	30°	22°

Figure 3.55 *Chisel cutting angles*

Files are classified by the following features:

- length
- kind of cut
- grade of cut (roughness)
- profile
- cross-sectional shape or most common use.

The grades of cut are: rough, bastard, second, smooth and dead smooth. These cuts vary with the length of a file. For example a short, second cut file will be smoother than a longer, smooth file. For further information on file cuts, see Table 3.8. The profiles and cross-sectional shapes of some typical files are shown in Figure 3.57.

Figure 3.58 shows how a file should be held and used. To file flat is very difficult and the skill only comes with years of continual practice. Cross filing is used for rapid material removal. Draw filing is only a finishing operation to improve the surface finish. It removes less metal per stroke than cross filing and can produce a hollow

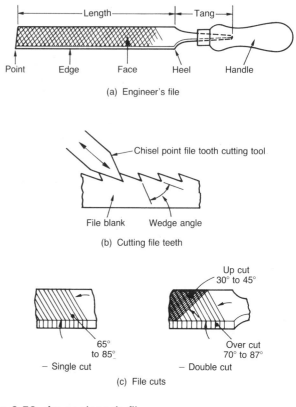

(a) Engineer's file

(b) Cutting file teeth

(c) File cuts

Figure 3.56 *An engineer's file*

Table 3.8 *Types of file*

Increase in pitch				
	Rough		Used for rapid metal removal. Not suitable if a good surface finish is required. Can be used on soft materials, the coarse pitch reduces clogging	The pitch of a file will increase as the length gets longer. For example the pitch of a 300mm second cut file is a larger than the pitch of a 150mm second cut file
	Bastard			
	Second cut		General purpose roughing and finishing	
	Smooth		Used when good surface finish and accurate dimensions are specified. Metal removal rate is poor	
	Dead smooth			

surface, unless care is taken.

The spaces between the teeth of a file tend to become clogged with bits of metal. This happens mostly when filing soft metals. It is called 'pinning'. The clogged teeth tend to leave heavy score marks in the surface of the work. These marks are difficult to remove. The file should be kept clean and a little chalk should be rubbed into the teeth to prevent pinning. Files are cleaned with a file brush called a 'file card'.

Test your knowledge 3.43

Name FOUR important safety precautions you should take when using a cold chisel.

Test your knowledge 3.44

Explain why a cold chisel must be held at the correct angle of inclination to the work.

Test your knowledge 3.45

Explain what is meant by a file having a 'safe edge' and name the more common types of file that have a 'safe edge'.

Test your knowledge 3.46

State how the cut of a file varies with its length.

Test your knowledge 3.47

Describe how you would produce a datum edge on a sawn blank by filing. The metal is 10 mm wide by 150 mm long.

	Type of file	Applications
□	Square	Filing of keyways and slots
△	Three square	Filing of angled surfaces
▽	Knife	Filing of acute angles
▯	Hand	These two files are the general-purpose tools for filing flat surfaces and convex profiles
▯	Flat	
○	Round	Used for enlarging or elongating holes
◗	Half round	Filing of concave profiles

Figure 3.57 *Types of file and applications*

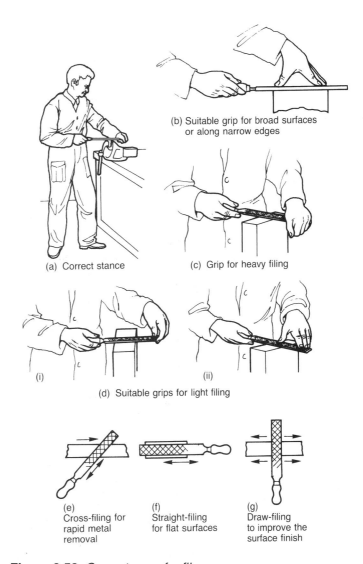

(a) Correct stance

(b) Suitable grip for broad surfaces or along narrow edges

(c) Grip for heavy filing

(i)　　(ii)

(d) Suitable grips for light filing

(e) Cross-filing for rapid metal removal

(f) Straight-filing for flat surfaces

(g) Draw-filing to improve the surface finish

Figure 3.58 *Correct use of a file*

Hacksaws and sawing

A typical hacksaw frame and blade is shown in Figure 3.59(a). The frame is adjustable so that it can be used with blades of various lengths. It is also designed to hold the blade in tension when the wing nut is tightened. The blade is put into the frame so that the teeth cut on the forward stroke. Figure 3.59(b) shows how a hacksaw should be held when being used.

There are a variety of blade types available:

- High speed steel 'all hard' blades are the most rigid and give the most accurate cut. However they are brittle and easily broken when used by an inexperienced person.
- High speed steel 'soft back' blades have a good life and, being more flexible, are less easily broken.
- Carbon steel flexible blades are satisfactory for occasional use on soft non-ferrous metals. They are cheap and not easily broken. Unfortunately they only have a limited life when cutting steels.

To prevent the blade from jamming in the slot it makes as it cuts, all saw blades are given a 'set'. This is shown in Figure 3.60. Coarse pitch hacksaw blades and power-saw blades have the individual teeth set to the left and to the right with either the intermediate teeth or every third tooth left straight to clear the slot. This as shown in Figure 3.60(a). This is not possible with fine pitch blades, and the blade as a whole is given a 'wave' set as shown in Figure 3.60(b). The effect of set on the cut being made is shown in Figure 3.60(c). If

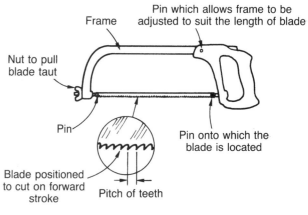

(a) Metal cutting hacksaw

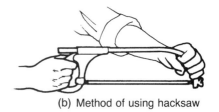

(b) Method of using hacksaw

Figure 3.59 *A typical hacksaw frame and blade*

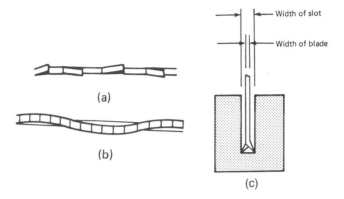

Figure 3.60 *The 'set' of a hacksaw blade*

you have to change a blade part way through a cut, never continue in the old slot. Because the set of the old blade will have worn, the new blade will jam in the old cut and break. Always start a new cut to the side of the failed cut.

The sizes of hacksaw blades are now given in metric sizes. The length (between the fixing hole centres), the width and the thickness. However, the pitch of the teeth is still given as so many teeth per inch.

The fewer teeth per inch the coarser will be the cut, the more quickly will the metal be removed, and the greater will be the set so that there is less chance of the blade jamming. However, there should always be at least three teeth in contact with the work at any one time. Therefore, the thinner the metal being cut the finer the pitch of the blade that should be used. Some typical examples are given in Table 3.9.

Table 3.9 *Hacksaw blade sizes and applications*

Teeth per inch	Material to be cut	Blade applications
32	Up to 3 mm	Thin sheets and tubes Hard and soft materials (thin sections)
24	3 mm to 6 mm	Thicker sheets and tubes Hard and soft materials (thicker sections)
18	6 mm to 12 mm	Heavier sections such as mild steel, cast iron, aluminium, brass, copper, bronze
14	Greater than 12 mm	Soft materials (such as aluminium, brass, copper, bronze) with thicker sections

Screw thread cutting

Internal screw threads are cut with taps. A set of straight fluted hand taps are shown in Figure 3.61. The difference between them is the length of the lead. The taper tap should be used first to start the thread. Great care must be taken to ensure that the tap is upright in the hole and it should be checked with a try square. The second tap is used to increase the length of thread and can be used for finishing if the tap passes through the work. The third tap is used for 'bottoming' in blind holes.

The hole to be threaded is called a 'tapping size' hole and it is the same size or only very slightly larger than the core diameter of the thread. Drill diameters for drilling tapping size holes for different screw threads can be found in sets of workshop tables. For example, the tapping size drill for an M10 × 1.5 thread is 8.5 mm diameter.

A tap wrench is used to rotate the taps. There are a variety of different styles available depending upon the size of the taps. An example of a suitable wrench for small taps is shown in Figure 3.62. Taps are very fragile and are easily broken, particularly in the small sizes. Once a tap has been broken into a hole, it is virtually impossible to get it out without damaging or destroying the workpiece.

Taps are relatively expensive and should be looked after carefully. High speed steel ground thread taps are the most expensive. However, they cut very accurate threads and, with careful use, last a long time. Carbon steel cut thread taps are less accurate and less expensive and have a reasonable life when cutting the softer non-ferrous metals. Whichever sort of taps are used, they should always be well lubricated. Traditionally tallow was used, but nowadays proprietary screw-cutting lubricants are available that are more effective.

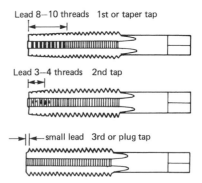

Figure 3.61 *Hand taps*

Figure 3.62 *A tap wrench*

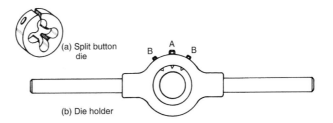

Figure 3.63 *A die holder*

External threads are cut using split button dies in a die holder, as shown in Figure 3.63. One face of the die is always marked up with details of the thread and the maker's logo. This should be visible when the die is in the die-holder. Then the lead is on the correct side for starting the cut. Screw A is used to spread the die for the first cut. The screws marked B are used to close the die until it gives the correct finishing cut. This is judged by using a standard nut or a screw thread gauge. The nut or gauge should run up the thread without binding or without undue looseness.

Again, the die must be started square with the workpiece or a 'drunken' thread will result. Also, a thread cutting lubricant should be used. Like thread cutting taps, dies are available in carbon steel cut thread and high speed steel ground thread types. For both taps and dies, each set only cuts one size and pitch of thread and one thread form.

Spanners and keys

In addition to cutting tools a fitter should also have a selection of spanners and keys available for dismantling and assembly purposes. Figure 3.64 shows a selection of spanners and keys. These are carefully proportioned so that a person of average strength will be able to tighten a screwed fastening correctly.

Use of a piece of tubing to extend a spanner or key is very bad practice. It strains the jaws of the spanner so that it becomes loose and may slip. It may even crack the jaws of the spanner so that they break. In both cases this can lead to nasty injuries to your hands and even a serious fall if you are working on a ladder. Also it over

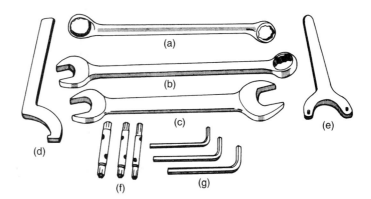

Figure 3.64 *A selection of spanners and keys*

stresses the fastening which will be weakened or even broken. Always check a spanner for damage and correct fit before using it. A torque spanner should be used to tighten important fastenings.

Drilling

Drilling is a process for producing holes. The holes may be cut from the solid or existing holes may be enlarged. The purpose of the drilling machine is to:

*   Rotate the drill at a suitable speed for the material being cut and the diameter of the drill.
*   Feed the drill into the workpiece.
*   Support the workpiece being drilled; usually at right-angles to the axis of the drill. On some machines the table may be tilted to allow holes to be drilled at a pre-set angle.

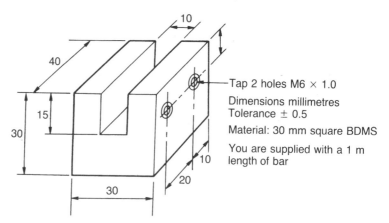

Tap 2 holes M6 × 1.0

Dimensions millimetres
Tolerance ± 0.5

Material: 30 mm square BDMS

You are supplied with a 1 m length of bar

Figure 3.65 *See Activity 3.10*

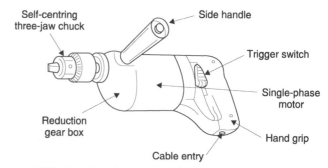

Figure 3.66 *An electric power drill*

Drilling machines

Drilling machines come in a variety of types and sizes. Figure 3.66 shows a hand held, electrically driven, power drill. It depends upon the skill of the operator to ensure that the drill cuts at right-angles to the workpiece. The feed force is also limited to the muscular strength of the user. Figure 3.67 shows a more powerful, floor mounted machine. The spindle rotates the drill. It can also move up and down in order to feed the drill into the workpiece and withdraw the drill at

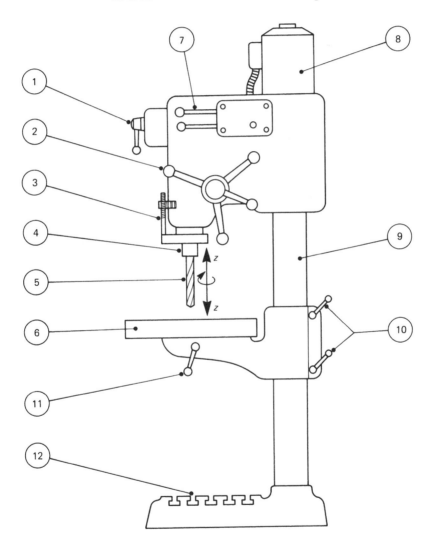

Parts of the Pillar Type Drilling Machine

1 Stop/start switch (electrics).

2 Hand or automatic feed lever.

3 Drill depth stop.

4 Spindle.

5 Drill.

6 Table.

7 Speed change levers.

8 Motor.

9 Pillar.

10 Vertical table lock.

11 Table lock.

12 Base.

Figure 3.67 *A pillar drill*

the end of the cut. Holes are generally produced with twist drills. Figure 3.68 shows a typical straight shank drill and a typical taper shank drill and names their more important features.

Large drills have taper shanks and are inserted directly into the spindle of the machine, as shown in Figure 3.69(a). They are located and driven by a taper. The tang of the drill is for extraction purposes only. It does not drive the drill. The use of a drift to remove the drill is shown in Figure 3.69(b).

Small drills have straight (parallel) shanks and are usually held in a self-centring chuck. Such a chuck is shown in Figure 3.69(c). The chuck is tightened with the chuck key shown. SAFETY: The chuck key must be removed before starting the machine. The drill chuck has a taper shank which is located in, and driven by, the taper bore of the drilling machine spindle.

The cutting edge of a twist drill is wedge-shaped, like all the tools we have considered so far. This is shown in Figure 3.70.

When regrinding a drill it is essential that the point angles are correct. The angles for general purpose drilling are shown in Figure 3.71(a). After grinding, the angles and lip lengths must be checked as shown in Figure 3.71(b). The point must be symmetrical. The effects of incorrect grinding are shown in Figure 3.71(c).

If the lip lengths are unequal, an oversize hole will be drilled when cutting from the solid. If the angles are unequal, then only one lip will cut and undue wear will result. The unbalanced forces will cause

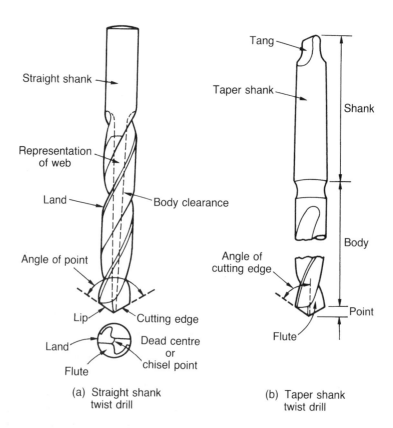

(a) Straight shank twist drill

(b) Taper shank twist drill

Figure 3.68 *Straight shank and taper shank twist drills*

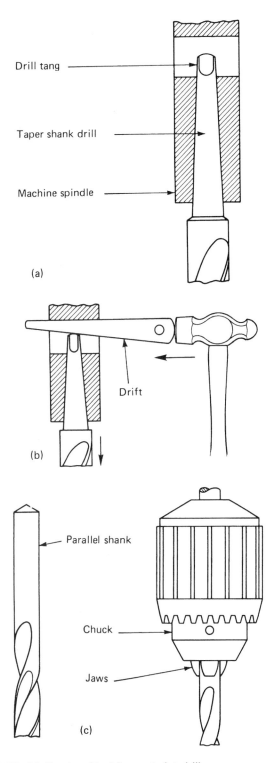

Drill tang

Taper shank drill

Machine spindle

(a)

Drift

(b)

Parallel shank

Chuck

Jaws

(c)

Figure 3.69 *Methods of holding a twist drill*

the drill to flex and 'wander'. The axis of the hole will become displaced as drilling proceeds. If both these faults are present at the same time, both sets of faults will be present and an inaccurate and ragged hole will result.

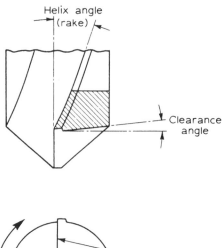

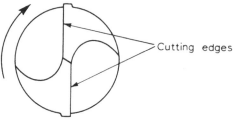

Figure 3.70 *Cutting edges of a twist drill*

Work-holding when drilling

It is dangerous to hold work being drilled by hand. There is always a tendency for the drill to grab the work and spin it round. Also the rapidly spinning *swarf* can produce some nasty cuts to the back of your hand. Therefore the work should always be securely fastened to the machine table.

Nevertheless, small holes in relatively large components are sometime drilled with the work handheld. In this case a stop bolted to the machine table should be used to prevent rotation.

Small work is usually held in a machine vice which, in turn, is securely bolted to the machine table. This is shown in Figure 3.72(a).

Larger work can be clamped directly to the machine table, as shown in Figure 3.72(b). In both these latter two examples the work is supported on parallel blocks. You mount the work in this way so that when the drill 'breaks through' the workpiece it does not damage the vice or the machine table.

Figure 3.72(c) shows how an angle plate can be used when the hole axis has to be parallel to the datum surface of the work. Finally, Figures 3.73(a) and 3.73(b) show how cylindrical work is located and supported using vee-blocks.

Miscellaneous drilling operations

Figure 3.74 shows some miscellaneous operations that are frequently carried out on drilling machines.

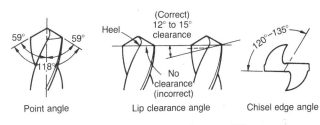

(a) Drill angles for general purpose drilling

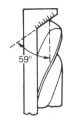

(b) Checking for correct point angle and equal lip lengths

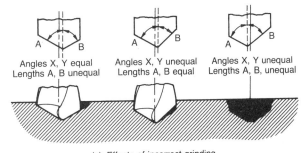

Angles X, Y equal Lengths A, B unequal   Angles X, Y unequal Lengths A, B equal   Angles X, Y unequal Lengths A, B, unequal

(c) Effects of incorrect grinding

Figure 3.71 *Point angles for a twist drill*

Countersinking

Figure 3.74(a) shows a countersink bit being used to countersink a hole to receive the heads of rivets or screws. For this reason the included angle is 90°. Lathe centre drills are unsuitable for this operation as their angle is 60°.

Counterboring

Figure 3.74(b) shows a piloted counterbore being used to counterbore a hole so that the head of a capscrew or a cheese-head screw can lie below the surface of the work. Unlike a countersink cutter, a counterbore is not self-centring. It has to have a pilot which runs in the previously drilled bolt or screw hole. This keeps the counterbore cutting concentrically with the original hole.

Spot-facing

This is similar to counterboring but the cut is not as deep. It is used to provide a flat surface on a casting or a forging for a nut and washer to 'seat' on. Sometimes, as shown in Figure 3.74(c), it is used to machine a boss (raised seating) to provide a flat surface for a nut and washer to 'seat' on.

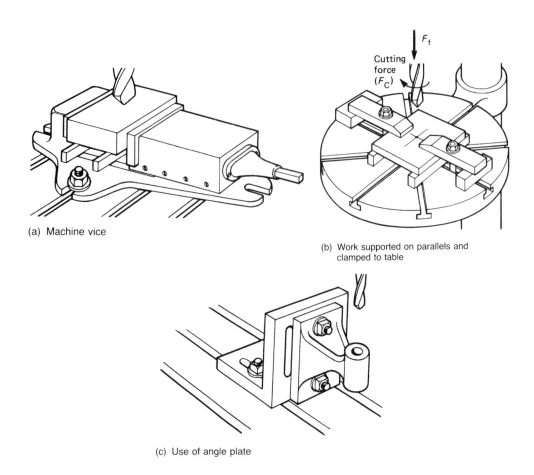

(a) Machine vice

(b) Work supported on parallels and clamped to table

(c) Use of angle plate

Figure 3.72 *Work-holding*

Centre lathe turning

The main purpose of a centre lathe is to produce external and internal cylindrical and conical (tapered) surfaces. It can also produce plain surfaces and screw threads.

The centre lathe

Figure 3.75(a) shows a typical centre lathe and identifies its more important parts.

- The bed is the base of the machine to which all the other sub-assemblies are attached. Slideways accurately machined on its top surface provide guidance for the saddle and the tail stock. These slideways also locate the headstock so that the axis of the spindle is parallel with the movement of the saddle and the tailstock. The saddle or carriage of the lathe moves parallel to the spindle axis as shown in Figure 3.75(b).
- The cross-slide is mounted on the saddle of the lathe. It moves at 90° to the axis of the spindle, as shown in Figure 3.75(c). It provides in-feed for the cutting tool when cylindrically turning. It

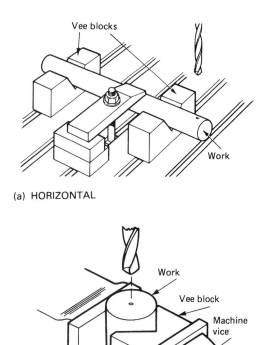

(a) HORIZONTAL

(b) VERTICAL

Figure 3.73 *Work-holding cylindrical components*

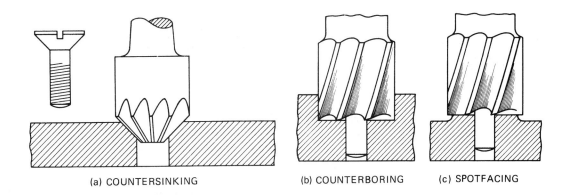

(a) COUNTERSINKING   (b) COUNTERBORING   (c) SPOTFACING

Figure 3.74 *Countersinking, counterboring and spot facing*

is also used to produce a plain surface when facing across the end of a bar or component.

• The top-slide (compound-slide) is used to provide in-feed for the tool when facing. It can also be set at an angle to the spindle axis for turning tapers, as shown in Figure 3.75(b).

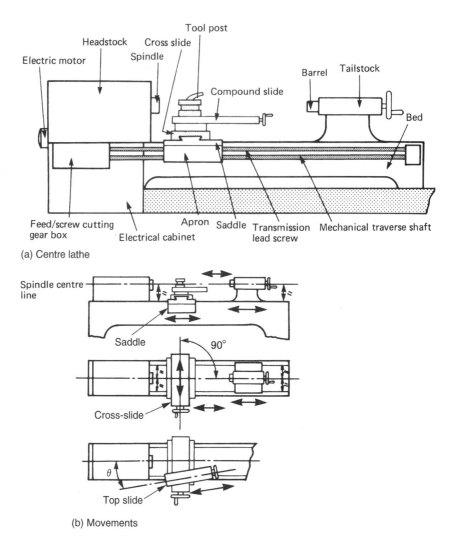

Figure 3.75 *A centre lathe*

Table 3.10 *Centre lathe movements*

Cutting movement	Hand or power traverse	Means by which movement is achieved	Turned feature
Tool parallel to the spindle centre line	Both	The saddle moves along the bed slideways	A parallel cylinder
Tool at 90° to the spindle centre line	Both	The cross slide moves along a slideway machined on the top of the saddle	A flat face square to the spindle centre line
Tool at an angle relative to the spindle centre line	Hand	The compound slide is rotated and set at the desired angle relative to the centre line	A tapered cone

Work-holding in the lathe

The work to be turned can be held in various ways. We will now consider the more important of these.

Between centres

The centre lathe derives its name from this basic method of work-holding. The general layout is shown in Figure 3.76(a). Centre holes are drilled in the ends of the bar and these locate on centres in the headstock spindle and the tailstock barrel. A section through a correctly centred component is shown in Figure 3.76(b). The centre-hole is cut with a standard centre-drill. The main disadvantage of this

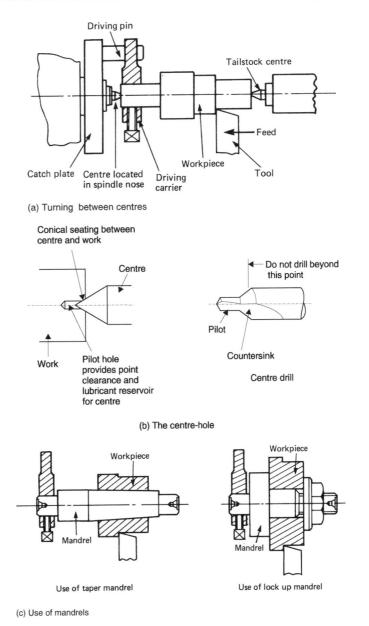

(a) Turning between centres

(b) The centre-hole

(c) Use of mandrels

Figure 3.76 *Work-holding in a centre lathe*

method of workholding is that no work can be performed on the end of the component. Work that has been previously bored can be finish turned between centres using a taper mandrel as shown in Figure 3.76(c).

Four-jaw chuck

Chucks are mounted directly onto the spindle nose and hold the work securely without the need for a back centre. This allows the end of the work to be faced flat. It also allows for the work to have holes bored into it or through it.

In the four-jaw chuck, the jaws can be moved independently by means of jack-screws. As shown in Figure 3.77(a), the jaws can also be reversed and the work held in various ways, as shown in Figure 3.77(b). As well as cylindrical work, rectangular work can also be held, as shown in Figure 3.77(c). Because the jaws can be moved independently, the work can be set to run concentrically with the spindle axis to a high degree of accuracy. Alternatively the work can be deliberately set off-centre to produce eccentric components as shown in Figure 3.77(d).

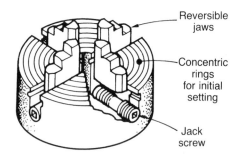

(a) Independent four-jaw chuck

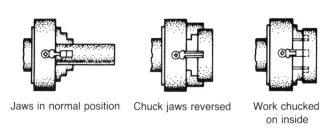

Jaws in normal position   Chuck jaws reversed   Work chucked on inside

(b) Methods of holding work in a chuck

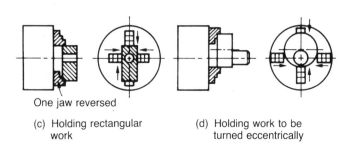

One jaw reversed

(c) Holding rectangular work

(d) Holding work to be turned eccentrically

Figure 3.77 *Four-jaw chuck*

Three-jaw chuck

The self-centring, three-jaw chuck is shown in Figure 3.78(a). The jaws are set at 120° and are moved in or out simultaneously (at the same time) by a scroll when the key is turned. SAFETY: This key must be removed before starting the lathe or a serious accident can occur. When new and in good condition this type of chuck can hold cylindrical and hexagonal work concentric with the spindle axis to a high degree of accuracy. In this case the jaws are not reversible, so it is provided with separate internal and external jaws. In Figure 3.78(a) the internal jaws are shown in the chuck, and the external jaws are shown at the side of the chuck. Again the chuck is mounted directly on the spindle nose of the lathe.

Work to be turned between centres is usually held in a three- jaw chuck whilst the ends of the bar are faced flat and then centre drilled, as shown in Figure 3.78(b).

Face-plate

Figure 3.79 shows a component held on a face-plate so that the hole can be bored perpendicularly to the datum surface. This datum

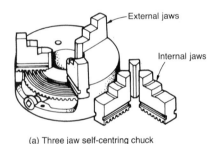

(a) Three jaw self-centring chuck

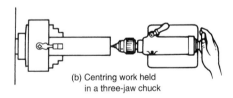

(b) Centring work held
in a three-jaw chuck

Figure 3.78 *Three-jaw chuck*

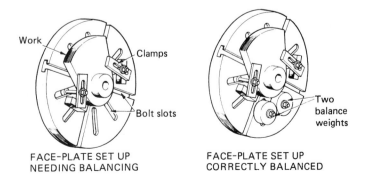

FACE-PLATE SET UP
NEEDING BALANCING

FACE-PLATE SET UP
CORRECTLY BALANCED

Figure 3.79 *Face-plate set up*

surface is in contact with the face-plate. Note that the face-plate has to be balanced to ensure smooth running. Care must be taken to check that the clamps will hold the work securely and do not foul the machine. The clamps must not only resist the cutting forces, but they must also prevent the rapidly rotating work from spinning out of the lathe.

Turning tools

Figure 3.80(a) shows a range of turning tools and some typical applications. Figure 3.80(b) shows how the metal-cutting wedge also applies to turning tools. Turning tools are fastened into a tool-post which is mounted on the top slide of the lathe. There are many different types of tool-post. The four-way turret tool-post shown in Figure 3.80(c) allows four tools to be mounted at any one time.

Parallel turning

Figure 3.81(a) shows a long bar held between centres. To ensure that the work is truly cylindrical with no taper, the axis of the tailstock centre must be in line with the axis of the headstock spindle. The saddle traverse provides movement of the tool parallel with the workpiece axis. You take a test cut and measure the diameter of the bar at both ends. If all is well, the diameter should be constant all along the bar. If not, the lateral movement of the tailstock needs to be adjusted until a constant measurement is obtained. The depth of cut is controlled by micrometer adjustment of the cross-slide.

Whilst facing and centre drilling the end of a long bar, a fixed steady is used. This supports the end of the bar remote from the chuck. A fixed steady is shown in Figure 3.81(b).

If the work is long and slender it sometimes tries to kick away from the turning tool or even climb over the tool. To prevent this happening a travelling steady is used. This is bolted to the saddle opposite to the tool, as shown in Figure 3.81(c).

Surfacing

A surfacing (facing or perpendicular-turning) operation on a workpiece held in a chuck is shown in Figure 3.82. The saddle is clamped to the bed of the lathe and the tool motion is controlled by the cross-slide. This ensures that the tool moves in a path at right-angles to the workpiece axis and produces a plain surface. In-feed of the cutting tool is controlled by micrometer adjustment of the top-slide.

Boring

Figure 3.83 shows how a drilled hole can be opened up using a boring tool. The workpiece is held in a chuck and the tool movement

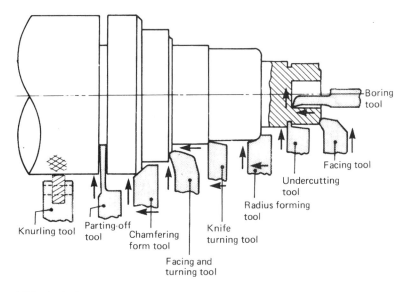

(a) Turning tools

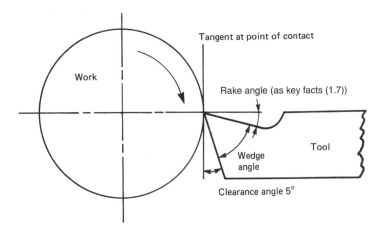

(b) Turning tool angles

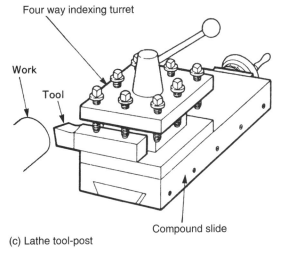

(c) Lathe tool-post

Figure 3.80 *Turning tools*

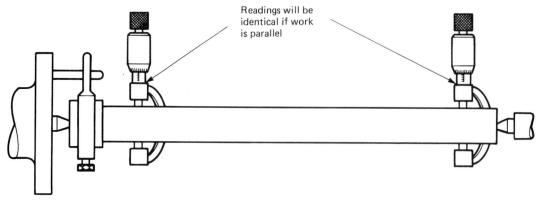

a) Checking for parallelism

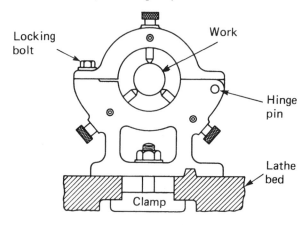

(b) Fixed steady

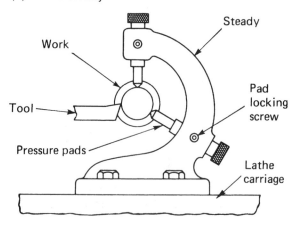

(c) Travelling steady

Figure 3.81 *Parallel turning*

is controlled by the saddle of the lathe. The in-feed of the tool is controlled by micrometer adjustment of the cross-slide. The pilot hole is produced either by a taper shank drill mounted directly into the tailstock barrel (poppet), or by a parallel shank drill held in a drill chuck. The taper mandrel of the drill chuck is inserted into the tailstock barrel.

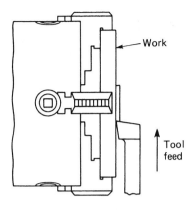

Figure 3.82 *Surfacing*

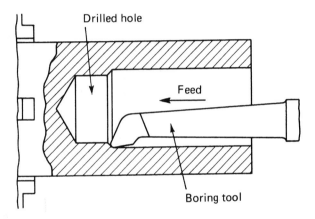

Figure 3.83 *Boring*

Conical surfaces

Chamfers on the corners of a turned component are short conical surfaces. These are usually produced by using a chamfering tool, as shown in Figure 3.80(a). Longer tapers can be produced by use of the top-slide. Use of the top (compound) slide is shown in Figure 3.84. The slide is mounted on a swivel base and it is fitted with a protractor scale. It can be swung round to the required angle and clamped in position. The taper is then cut as shown.

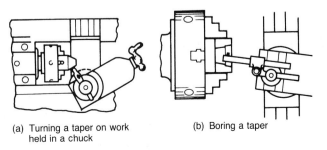

(a) Turning a taper on work held in a chuck

(b) Boring a taper

Figure 3.84 *Producing a chamfer*

Miscellaneous turning operations

Reamers are sizing tools. They remove very little metal. Since they follow the existing hole, they cannot correct the positional errors. Hand reamers have a square on the end of their shanks so that they can be rotated by a tap wrench. Machine reamers have a standard morse taper shank.

Figure 3.85(a) shows a hole being reamed in a lathe. A machine reamer is being used and it is held in the barrel of the tailstock. Because a drilled hole invariably runs out slightly, the pilot hole should be single-point bored in order to correct the position and geometry of the hole. It is finally sized using the reamer. Only the minimum amount of metal for the surface of the hole to clean up should be left in for the reamer to remove.

Standard, non-standard and large diameter screw threads can be cut in a centre lathe by use of the lead-screw to control the saddle movement. This is a highly skilled operation.

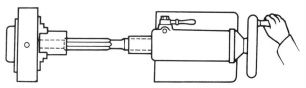

(a) Machine reamer supported in the tailstock

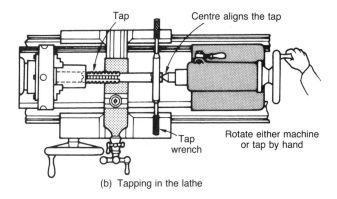

(b) Tapping in the lathe

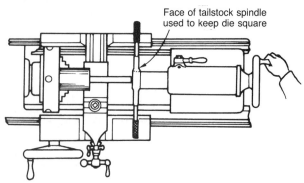

(c) Threading in the lathe with hand dies

Figure 3.85 *Miscellaneous turning operations*

However, standard screw threads of limited diameter can be cut using hand threading tools as shown in Figures 3.85(b) and 3.85(c). Taps are very fragile and the workpiece should be rotated by hand with the lathe switched off and the gears disengaged.

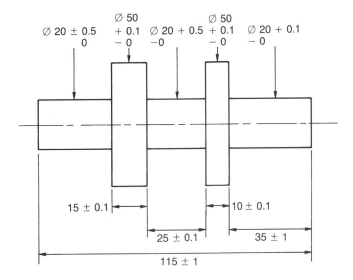

Dimensions in millimetres
Material: Phosphor bronze: Blank size ∅ 60 × 125
Hint: Turn between centres, after facing and centring

(a)

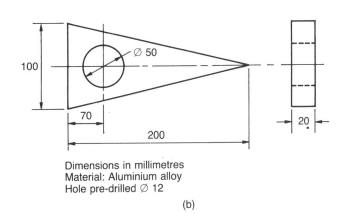

Dimensions in millimetres
Material: Aluminium alloy
Hole pre-drilled ∅ 12

(b)

Figure 3.86 *See Activity 3.11 and Test your knowledge 3.59*

Activity 3.11

Select suitable tools and measuring equipment (giving reasons for your choice) and draw up a production plan (see Page 116) for making the component shown in Figure 3.86. Present your work in the form of a word processed report with hand drawn sketches.

Joining and assembly

Engineered products usually comprise a number of components that must be assembled and joined together in some particular way. This section introduces the main methods used.

Assembly

The purpose of assembly is to put together a number of individual components to build up a whole device, structure or system. To achieve this aim attention must be paid to the following key factors.

Sequence of assembly

This must be planned so that as each component is added, its position in the assembly and the position of its fastenings are accessible. Also the sequence of assembly must be planned so that the installation of one component does not prevent access for fitting the next component or some later component.

Technique of joining

These must be selected to suit the components being joined, the materials from which they are made, and what they do in service. If the joining technique involves heating, then care must be taken that adjacent components are not heat sensitive or flammable.

Position of joints

Joints must not only be accessible for initial assembly, they must also be accessible for maintenance. You don't want to dismantle half a machine to make a small adjustment, or replace a part that wears out regularly.

Interrelationship and identification of parts

Identification of parts and their position in an assembly can usually be determined from assembly drawings or exploded view drawings. Interrelationship markings are often included on components. For example, the various members and joints of structural steelwork are given number and letter codes to help identification on site. Printed circuit boards usually have the outline of the various components printed on them as well as the part number.

Tolerances

The assembly technique must take into account the accuracy and finish of the components being assembled. Much greater care has to be taken when assembling a precision machine tool or an artificial satellite, than when assembling structural steel work.

Protection of parts

Components awaiting assembly require protection against accidental damage and corrosion. In the case of structural steelwork this may merely consist of painting with red oxide primer and careful stacking. Precision components will require treating with an anti-corrosion lanolin based compound that can be easily removed at the time of assembly. Bores must be sealed with plastic plugs and screw threads with plastic caps. Precision ground surfaces must also be protected from damage. Heavy components must be provided with eye-bolts for lifting. Vulnerable sub-assemblies such as aircraft engines must be supported in suitable cradles.

Joining (mechanical)

The joints used in engineering assemblies may be divided into the following categories.

Permanent joints
These are joints in which one or more of the components and/or the joining medium has to be destroyed or damaged in order to dismantle the assembly; for example, a riveted joint.

Temporary joints
These are joints that can be dismantled without damage to the components. It should be possible to re-assemble the components using the original or new fastenings; for example, a bolted joint.

Flexible joints
These are joints in which one component can be moved relative to another component in an assembly in a controlled manner; for example, the use of a hinge.

Screwed fastenings

These are used to make temporary joints that can be dismantled and re-assembled at will. They are required where maintenance is necessary. We considered different types of screwed fastenings in the section on component selection. We also considered the different types of head found on screwed fastenings. Figure 3.87 shows the correct way to use some typical screwed fastenings.

Screwed fastenings must always pull down onto prepared seatings that are flat and at right-angles to the axis of the fastening. This prevents the bolt or screw being bent as it is tightened up. To protect the seating, a soft washer is placed between the seating and the nut. Taper washers are used when erecting steel girders to prevent the draught angle of the flanges from bending the bolt.

Locking devices are used to prevent screwed fastenings from slackening off due to vibration. Locking devices may be frictional or positive. A selection of plain washers, taper washers and locking devices is shown in Figure 3.88.

Riveting

The selection of rivets and rivet heads was considered in the section on component selection. To make a satisfactory riveted joint the following points must be observed.

Hole clearance
If the clearance is too small there will be difficulty in inserting the

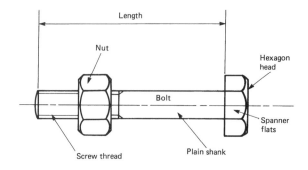

(a) HEXAGON HEAD BOLT AND NUT

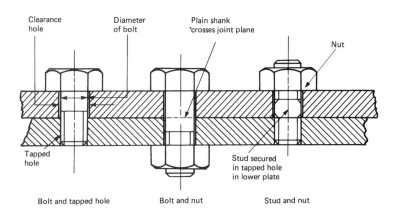

(b) TYPES OF SCREWED JOINT

Figure 3.87 *Screwed fastenings*

rivet and drawing up the joint. If the hole is too large, the rivet will buckle and a weak joint will result.

Rivet length

If the rivet is too long, the rivet bends over during heading. If the rivet is too short the head cannot be properly formed. In either case a weak joint will result. Figure 3.89(a) shows the correct proportions for a riveted joint and Figure 3.89(b) shows some typical riveting faults.

The correct procedure for heading (closing) a rivet is shown in Figure 3.90(a). The drawing up tool ensures that the components to be drawn are brought into close contact and that the head of the rivet is drawn up tightly against the lower component. The hammer blows with the flat face of the hammer swells the rivet so that it is a tight fit in the hole and starts to form the head. The ball pein of the hammer head is then used to rough form the rivet head. The head is finally finished and made smooth by using a rivet snap. Where large rivets and large quantities of rivets are to be closed a portable pneumatic

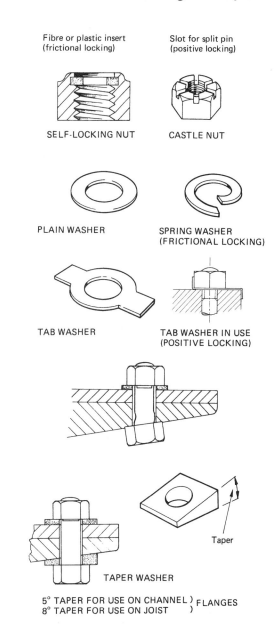

Fibre or plastic insert
(frictional locking)

Slot for split pin
(positive locking)

SELF-LOCKING NUT

CASTLE NUT

PLAIN WASHER

SPRING WASHER
(FRICTIONAL LOCKING)

TAB WASHER

TAB WASHER IN USE
(POSITIVE LOCKING)

Taper

TAPER WASHER

5° TAPER FOR USE ON CHANNEL) FLANGES
8° TAPER FOR USE ON JOIST)

Figure 3.88 *Various nuts and washers*

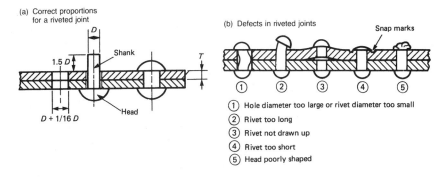

(a) Correct proportions
for a riveted joint

D

Shank

1.5 D

T

D + 1/16 D

Head

(b) Defects in riveted joints

Snap marks

① ② ③ ④ ⑤

① Hole diameter too large or rivet diameter too small
② Rivet too long
③ Rivet not drawn up
④ Rivet too short
⑤ Head poorly shaped

Figure 3.89 *Correct and incorrect riveted joints*

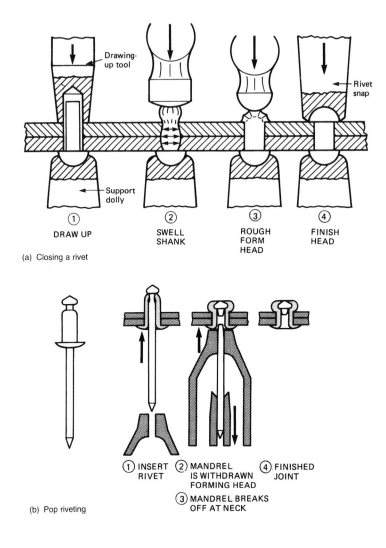

Figure 3.90 *Correct procedure for riveting*

tool is used.

'Pop' riveting is often used for joining thin sheet metal components, particularly when building up box sections. When building up box sections it is not possible to get inside the closed box to use a hold-up dolly; for example, when riveting the skin to an aircraft wing. Figure 3.90(b) shows the principle of 'pop' riveting.

Fusion welding

Welding has largely taken over from riveting for many purposes such as ship and bridge building and for structural steelwork. Welded joints are continuous and, therefore, transmit the stresses across the joint uniformly. In riveted joints the stresses are concentrated at each rivet. Also the rivet holes reduce the cross-sectional areas of the members being joined and weaken them. However, welding is a

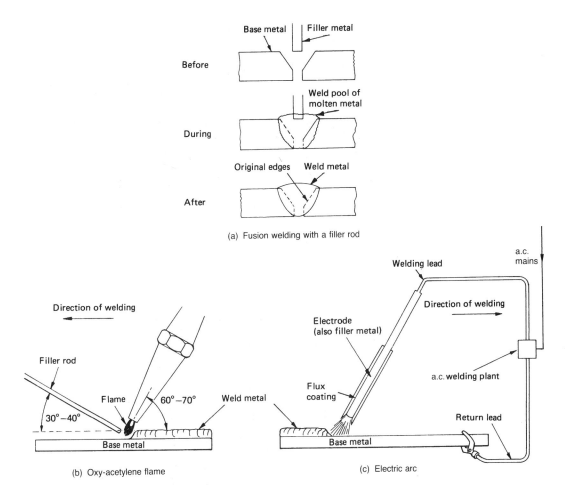

Base metal | Filler metal

Before

Weld pool of
molten metal

During

Original edges | Weld metal

After

(a) Fusion welding with a filler rod

Direction of welding

Filler rod

Flame 60°–70° Weld metal

30°–40°

Base metal

(b) Oxy-acetylene flame

Welding lead

a.c. mains

Direction of welding

Electrode
(also filler metal)

a.c. welding plant

Flux
coating

Return lead

Base metal

(c) Electric arc

Figure 3.91 *Fusion welding*

more skilled assembly technique and the equipment required is more costly. The components being joined are melted at their edges and additional filler metal is melted into the joint. The filler metal is of similar composition to that of the components being joined. Figure 3.91(a) shows the principle of fusion welding.

High temperatures are involved to melt the metal of the components being joined. These can be achieved by using the flame of an oxy-acetylene blowpipe, as shown in Figure 3.91(b), or an electric arc, as shown in Figure 3.91(c). When oxy-acetylene welding (gas welding), a separate filler rod is used. When arc welding, the electrode is also the filler rod and is melted as welding proceeds.

No flux is required when oxy-acetylene welding as the molten metal is protected from atmospheric oxygen by the burnt gases (products of combustion). When arc welding, a flux is required. This is in the form of a coating surrounding the electrode. This flux coating is not only deposited on the weld to protect it, it also stabilizes the arc and makes the process easier. The hot flux gives off fumes and adequate ventilation is required.

Protective clothing must be worn when welding and goggles or a face mask (visor) appropriate for the process must be used. These have optical filters that protect the user's eyes from the harmful radiations produced during welding. The optical filters must match the process.

The compressed gases used in welding are very dangerous and welding equipment must only be used by skilled persons or under close supervision. Acetylene gas bottles must only be stored and used in an upright position.

The heated area of the weld is called the weld zone. Because of the high temperatures involved, the heat affected area can spread back into the parent metal of the component for some distance from the actual weld zone. This can alter the structure and properties of the material so as to weaken it and make it more brittle. If the joint fails in service, failure usually occurs at the side of the weld in this heat affected zone. The joint itself rarely fails.

Test your knowledge 3.60

State THREE advantages and THREE disadvantages of welding compared with riveting.

Soft soldering

Soft soldering is also a thermal jointing process. Unlike fusion welding, the parent metal is not melted and the filler metal is an alloy of tin and lead that melts at relatively low temperatures. Soft soldering is mainly used for making mechanical joints in copper and brass components (plumbing). It is also used to make permanent electrical connections. Low carbon steels can also be soldered providing the metal is first cleaned and then *tinned* using a suitable flux. The tin in the solder reacts chemically with the surface of the component to form a bond.

Figure 3.92 shows how to make a soft soldered joint. The surfaces to be joined are first degreased and physically cleaned to remove any dust and dirt. Fine abrasive cloth or steel wool can be used. A flux is used to render the joint surfaces chemically clean and to make the solder spread evenly through the joint. Some soft soldering fluxes and their typical applications are listed in Table 3.11.

- The copper *bit* of the soldering iron is then heated. For small components and fine electrical work an electrically heated iron can be used. For joints requiring a soldering iron with a larger bit, a gas heated soldering stove can be used to heat the bit.
- The heated bit is then cleaned, fluxed and coated with solder. This is called *tinning* the bit.
- The heated and tinned bit is drawn slowly along the fluxed surfaces of the components to be joined. This transfers solder to the surfaces of the components. Additional solder can be added if required. The work should be supported on wood to prevent heat loss. The solder does not just 'stick' to the surface of the metal being tinned. The solder reacts chemically with the surface to form an amalgam that penetrates into the surface of the metal. This forms a permanent bond.
- Finally the surfaces are overlapped and 'sweated' together. That is, the soldering iron is re-heated and drawn along the joint as shown. Downward pressure is applied at the same time. The solder in the joint melts. When it solidifies it forms a bond between the two components.

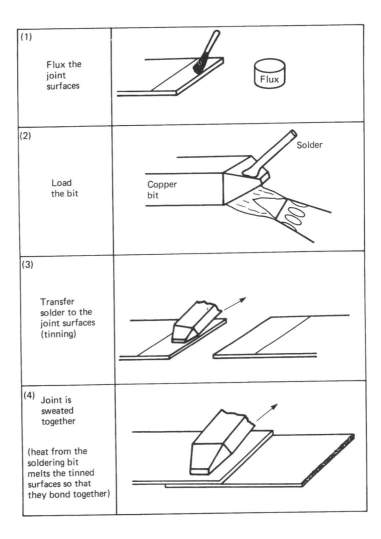

(1)	Flux the joint surfaces	
(2)	Load the bit	Solder / Copper bit
(3)	Transfer solder to the joint surfaces (tinning)	
(4)	Joint is sweated together (heat from the soldering bit melts the tinned surfaces so that they bond together)	

Figure 3.92 *Procedure for making a soft soldered joint*

Table 3.11 *Soldering fluxes*

Flux	Metals	Characteristics
Killed spirits (acidulated zinc chloride solution)	Steel, tin plate, brass and copper	Powerful cleansing action but leaves corrosive residue
Dilute hydrochloric acid	Zinc and galvanized iron	As above, wash after use
Resin paste or 'cored' solder	Electrical conductor and terminal materials	Only moderate cleansing action (passive flux) but non-corrosive
Tallow	Lead and pewter	As above
Olive oil	Tin plate	Non-toxic, passive flux for food containers, non-corrosive

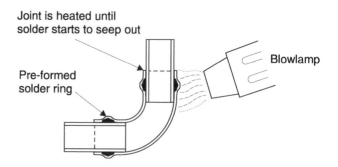

Figure 3.93 *'Sweating' a copper pipe*

Figure 3.93 shows how a copper pipe is sweated to a fitting. The pipe and the fitting are cleaned, fluxed and assembled. The joint is heated with a propane gas torch and solder is added. This is usually a resin-flux cored solder. The solder is drawn into the close fitting joint by capillary action.

Hard soldering

Hard soldering uses a solder whose main alloying elements are copper and silver. Hard soldering alloys have a much higher melting temperature range than soft solders. The melting range for a typical soft solder is 183–212°C. The melting range for a typical hard solder is 620–680°C. Hard soldering produces joints that are stronger and more ductile. The melting range for hard solders is very much lower than the melting point of copper and steel, but it is only just below the melting point of brass. Therefore great care is required when hard soldering brass to copper. Because the hard solder contains silver it is often referred to as 'silver solder'. A special flux is required based on borax.

A soldering iron cannot be used because of the high temperatures involved. Heating is by a blow pipe. Figure 3.94 shows you how to make a typical hard soldered joint. Again cleanliness and careful

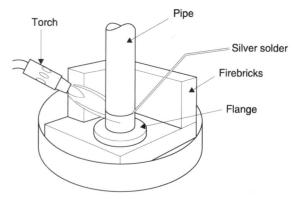

- The work is up to heat when the silver solder melts on contact with the work with the flame momentarily withdrawn.
- Add solder as required until joint is complete.

Figure 3.94 *Procedure for making a hard soldered joint*

List the advantages and limitations of soft soldering compared with hard soldering.

surface preparation is essential for a successful joint. The joint must be close fitting and free from voids. The silver solder is drawn into the joint by capillary action.

Even stronger joints can be made using a brass alloy instead of a silver–copper alloy. This is called *brazing*. The temperatures involved are higher than those for silver soldering. Therefore, brass cannot be brazed. The process of brazing is widely used for joining the steel tubes and malleable cast iron fittings of bicycle frames.

Activity 3.12

Consult manufacturers' or suppliers' data and draw up a table showing the composition of several soft solders and their typical applications. Present your work in the form of an overhead projector transparency.

Adhesive bonding

The advantages of adhesive bonding can be summarized as follows.

- The temperature rise from the curing of the adhesive is negligible compared with that of welding. Therefore the properties of the materials being joined are unaffected.
- Similar and dissimilar materials can be joined.
- Adhesives are electrical insulators. Therefore they reduce or prevent electrolytic corrosion when dissimilar metals are joined together.
- Joints are sealed and fluid tight.
- Stresses are transmitted across the joint uniformly.
- Depending upon the type of adhesive used, some bonded joints tend to damp out vibrations.

Bonded joints have to be specially designed to exploit the properties of the adhesive being used. You cannot just substitute an adhesive in a joint designed for welding, brazing or soldering. Figure 3.95(a) shows some typical bonded joint designs that provide a large contact area. A correctly designed bonded joint is very strong. Major structural members in modern high performance airliners and military aircraft are adhesive bonded. Figure 3.95(b) defines some of the jargon used when talking about bonded joints.

The strength of a bonded joint depends upon two factors.

Adhesion

This is the ability of the adhesive to 'stick' to the materials being joined (the *adherends*). This can result from physical keying or interlocking, as shown in Figure 3.96(a). Alternatively specific bonding can take place. Here, the adhesive reacts chemically with the surface of the adherends, as shown in Figure 3.96(b). Bonding occurs through intermolecular attraction.

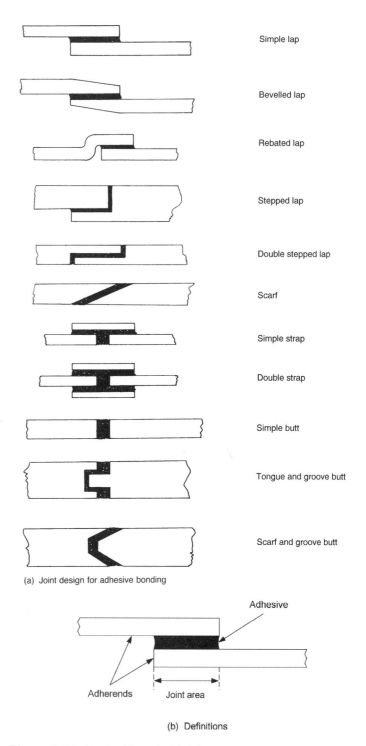

(a) Joint design for adhesive bonding

(b) Definitions

Figure 3.95 *Typical bonded joints*

Cohesion

This is the internal strength of the adhesive. It is the ability of the adhesive to withstand forces within itself. Figure 3.96(c) shows the failure of a joint made from an adhesive that is strong in adhesion but weak in cohesion. Figure 3.96(d) shows the failure of a joint that is

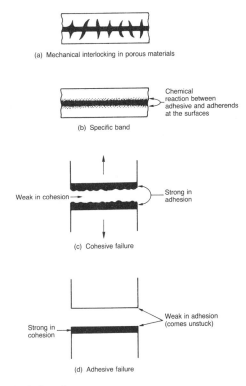

(a) Mechanical interlocking in porous materials

(b) Specific band

(c) Cohesive failure

(d) Adhesive failure

Figure 3.96 *Adhesion*

strong in cohesion but weak in adhesion.

As well as the design of the joint, the following factors affect the strength of a bonded joint:

- The joint must be physically clean and free from dust, dirt, moisture, oil and grease.
- The joint must be chemically clean. The materials being joined must be free from scale or oxide films.
- The environment in which bonding takes place must have the correct humidity and be at the correct temperature.

Bonded joints may fail in four ways. These are shown in Figure 3.97. Bonded joints are least likely to fail in tension and shear. They are most likely to fail in cleavage and peel.

The most efficient way to apply adhesives is by an adhesive gun. This enables the correct amount of adhesive to be applied to the correct place without wastage or mess. It also prevents the evaporation of highly flammable and toxic solvents whilst the adhesive is waiting to be used.

Joining (electrical and mechanical)

Again, joints may be permanent or temporary. Permanent joints are soldered or crimped. Temporary joints are bolted, clamped or plugged in.

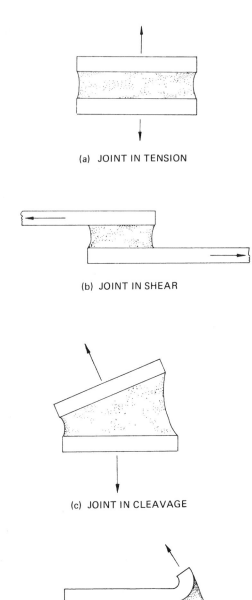

(a) JOINT IN TENSION

(b) JOINT IN SHEAR

(c) JOINT IN CLEAVAGE

(d) JOINT IN PEEL

Figure 3.97 *Ways in which bonded joints can fail*

Soldered joints

When soldering electrical and electronic components it is important not to overheat them. Overheating can soften thermoplastic insulation and completely destroy solid state devices such as diodes and transistors. Very often some form of heat sink is required when soldering solid state devices.

A high tin content low melting temperature solder with a resin flux core should be used. This is a passive flux. It only protects the joint.

It contains no active, corrosive chemicals to clean the joint. Therefore the joint must be kept clean whilst soldering. Even the natural grease from your fingers is sufficient to cause a high resistance 'dry' joint.

Figure 3.98(a) shows how a soldered connection is made to a solder tag. Note how the lead from the resistor is secured around the tag before soldering. This gives mechanical strength to the connection. Soldering provides the electrical continuity.

Figure 3.98(b) shows a prototype electronic circuit assembled on a matrix board. The board is made from laminated plastic and is pierced with a matrix of equally spaced holes. Pin tags are fastened into these holes in convenient places and the components are soldered to these pin tags.

Figure 3.98(c) shows the same circuit built up on a strip board. This is a laminated plastic board with copper tracks on the underside. The wire tails from the components pass through the holes in the board and are soldered to the tracks on the underside. The copper tracks are cut wherever a break in the circuit is required.

Figure 3.98(d) shows the underside of a printed circuit board (PCB). This is built up as shown in Figure 3.98(c), except that the tracks do not need to be cut since they are customized for the circuit.

Large volume assembly of printed circuit boards involves the use of pick and place robots to install the components. The assembled boards are then carried over a flow soldering tank on a conveyor. A roller rotates in the molten solder creating a 'hump' in the surface of the solder. As the assembled and fluxed board passes over this 'hump' of molten solder the components tags are soldered into place.

Wire wrapping

Wire wrapping is widely used in telecommunications where large numbers of fine conductors have to be terminated quickly and in close proximity to each other. Soldering would be inconvenient and the heat could damage the insulation of adjoining conductors. Also soldered joints would be difficult to disconnect. A special tool is used that automatically strips the insulation from the wire and binds the wire tightly around the terminal pins. The terminal pins are square in section with sharp corners. The corners cut into the conductor and prevents it from unwinding. The number of turns round the terminal is specified by the supervising engineer.

Crimped joints

For power circuits, particularly in the automotive industry, cable lugs and plugs are crimped onto the cables. The sleeve of the lug or the plug is slipped over the cable and then indented by a small pneumatic or hydraulic press. This is quicker than soldering and, as no heat is involved, there is no danger of damaging the insulation. Portable equipment is also available for making crimped joints on site. Hand operated equipment can be used to fasten lugs to small cables by crimping, as shown in Figure 3.99.

Clamped connections

Finally, we come to clamped connections using screwed fastenings. You will have seen many of these in domestic plugs, switches and lamp-holders. For heavier power installations, cable lugs are bolted to solid copper bus-bars using brass or bronze bolts, as shown in Figure 3.100.

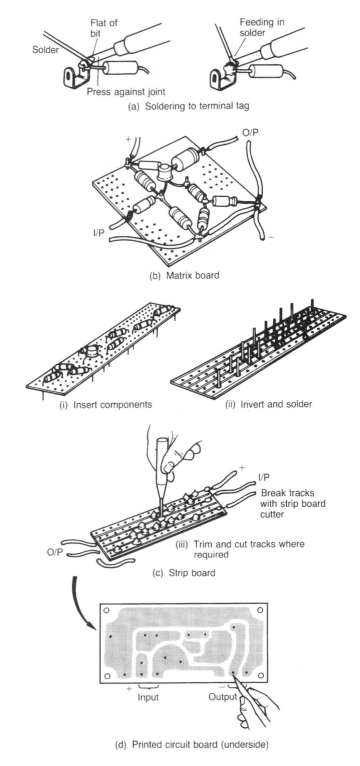

(a) Soldering to terminal tag

(b) Matrix board

(i) Insert components   (ii) Invert and solder

(iii) Trim and cut tracks where required

(c) Strip board

(d) Printed circuit board (underside)

Figure 3.98 *Various methods of electronic circuit assembly*

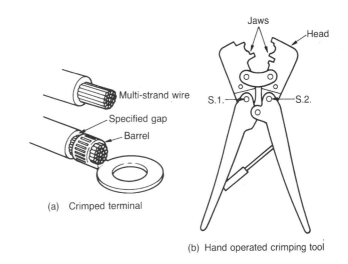

(a)   Crimped terminal

(b) Hand operated crimping tool

Figure 3.99 *Crimping*

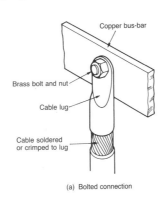

(a) Bolted connection

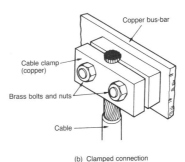

(b) Clamped connection

Figure 3.100 *Bolted and clamped connections*

Activity 3.14

A manufacturer of electronic kits has asked you to produce a single page instruction sheet on 'How to solder your kit'. Produce a word processed instruction sheet and illustrate it using appropriate sketches and drawings. Include a section headed 'Safety'.

Heat treatment

The properties of many metals and alloys can be changed by heating them to specified temperatures and cooling them under controlled conditions at specified rates. These are called, respectively, critical temperatures and critical cooling rates. We are only going to consider the heat treatment of plain carbon steels.

Hardening

The degree of hardness that can be given to any plain carbon steel depends upon two factors: the amount of carbon present, and how quickly the steel is cooled from the hardening temperature. The hardening temperature for medium carbon steels containing up to 0.8% carbon is bright red heat. Above 0.8% carbon the hardening temperature is dull red (cherry red) heat. Table 3.12(a) relates hardness to carbon content. Table 3.12(b) relates hardness to rate of cooling.

If oil quenching is used, a number of safety rules must be observed because of its flammability.

- Never use lubricating oil
- Always use a good quality quenching oil
- Always use a metal quenching bath with an airtight metal lid to smother the flames should the oil ignite

Table 3.12

(a) Effect of carbon content

Type of steel	Carbon content (%)	Effect of quench hardening
Low carbon	< 0.3	Negligible
Medium carbon	0.3–0.5 0.5–0.9	Becomes tougher Becomes hard
High carbon	0.9–1.3	Becomes very hard

(b) Rate of cooling

Carbon content (%)	Quenching media	Required treatment
0.3–0.5	Oil	Toughening
0.5–0.9	Oil	Toughening
0.5–0.9	Water	Hardening
0.9–1.2	Oil	Hardening

- Always keep the bath covered when not in use to keep out foreign objects and to avoid the absorption of moisture from the atmosphere.

Hardening faults

Under heating
If the temperature of a steel does not reach its critical temperature, the steel won't harden.

Overheating
It is a common mistake to think that increasing the temperature from which the steel is quenched will increase its hardness. Once the correct temperature has been reached, the hardness will depend only upon the carbon content of the steel and its rate of cooling. If the temperature of a steel exceeds its critical temperature, grain growth will occur and the steel will be weakened. Also overheating will slow the cooling rate and will actually reduce the hardness of the steel.

Cracking
There are many causes of hardening cracks. Some of the more important are: sharp corners, sudden changes of section, screw threads, holes too near the edge of a component. These should all be avoided at the design stage as should over rapid cooling for the type of steel being used.

Distortion
There are many causes of distortion. Some of the more important are as follows:

- lack of balance or symmetry in the shape of the component
- lack of uniform cooling. Long thin components should be dipped end-on into the quenching bath
- change in the grain structure of the steel causing shrinkage.

No matter how much care is taken when quench hardening, some distortion (movement) will occur. Also slight changes in the chemical composition may occur at the surface of the metal. Therefore, precision components should be finish ground after hardening. The components must be left slightly oversize before grinding to allow for this. That is, a grinding allowance must be left on such components before hardening.

Tempering

Quench-hardened plain carbon steels are very brittle and unsuitable for immediate use. Therefore, further heat treatment is required. This is called tempering. It greatly increases the toughness of the hardened steel at the expense of some loss of hardness.

Tempering consists of re-heating the hardened steel and again quenching it in oil or water. Typical tempering temperatures for various applications are summarized in Table 3.13. You can judge the tempering temperature by the colour of the oxide film. First, the component must be polished after hardening and before tempering. Then heat the component gently and watch for the colour of the metal surface to change. When you see the appropriate colour appear, the component must be quenched immediately.

Table 3.13 *Tempering temperatures*

Component	Temper colour	Temperature (°C)
Edge tools	Pale straw	220
Turning tools	Medium straw	230
Twist drills	Dark straw	240
Taps	Brown	250
Press tools	Brown/purple	260
Cold chisels	Purple	280
Springs	Blue	300
Toughening (medium carbon steels)	—	450–600

Chemical treatment

The chemical treatments that will be considered in this section are as follows:

- Chemical machining (etching) as used in the production of printed circuit boards
- The coating of metal components with decorative and/or corrosion resistant finishes.

Etching

Printed circuit boards have already been introduced in the section on assembly. First the circuit is drawn out by hand or designed using a computer aided design (CAD) package. A typical circuit master drawing is shown in Figure 3.101. The master drawing of the circuit is then photographed to produce a transparent copy called a negative. In a negative copy the light and dark areas are reversed.

The printed circuit board is made as follows:

- The copper coated laminated plastic (Tufnol) or fibre glass board is coated with a photoresist by dipping or spraying

- The negative of the circuit is placed in contact with the prepared circuit board. They are then exposed to ultraviolet light. The light passes through the transparent parts of the negative. The areas of the board exposed to the ultraviolet light will eventually become the circuit
- The exposed circuit board is then developed in a chemical solution that hardens the exposed areas
- The photoresist is stripped away from the unexposed areas of the circuit board
- The circuit board is then placed in a suitable etchant. Ferric chloride solution can be used as an etchant for copper. This eats away the copper where it is not protected by the hardened photoresist. The remaining copper is the required circuit
- The circuit boards are washed to stop the reaction. The remaining photoresist is removed so as not to interfere with the tinning of the circuit with soft solder and the soldering of the components into position.

SAFETY: This process is potentially dangerous. Ultraviolet light is very harmful to your skin and to your eyes. The ferric chloride solution is highly corrosive to your skin. The various processes also give off harmful fumes. Therefore you should only carry out this process under properly supervised, controlled and ventilated conditions. Appropriate protective clothing should be worn.

A similar process can be used for the chemical engraving of components with their identification numbers and other data.

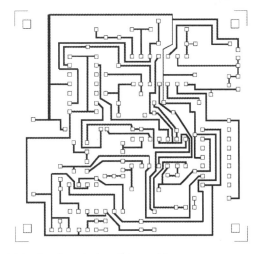

Figure 3.101 *A printed circuit board layout*

Electroplating

The component to be plated is placed into a plating bath as shown in Figure 3.102. The component is connected to the negative terminal of a direct current supply. This operates at a low voltage but relatively heavy current. The anode is connected to the positive terminal of the supply. The anode is usually made from the same metal as that which is to be plated onto the component. The electrolyte is a solution of

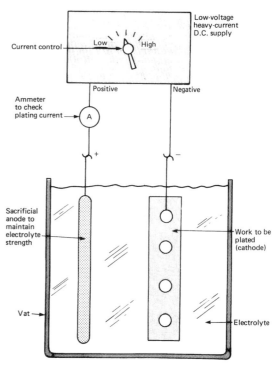

Figure 3.102 *Electroplating*

chemical salts. The composition depends upon the process being carried out.

When the current passes through the bath the component is coated with a thin layer of the protective and/or decorative metal. This metal comes from the chemicals in the electrolyte. The process is self-balancing. The anode dissolves into the electrolyte at the same rate as the metal taken from the electrolyte is being deposited on the component. This applies to most plating processes such as zinc, copper, tin and nickel plating.

An exception is chromium plating. A neutral anode is used that does not dissolve into the electrolyte. Additional salts have to be added to the electrolyte from time to time to maintain the solution strength. Chromium is not usually deposited directly onto the component. The component is usually nickel plated and polished and then a light film of chromium is plated over the nickel as a decorative and sealing coat.

Test your knowledge 3.69

Give TWO examples of where electroplating is used. Name the type of plating used in each case.

Electrolytic galvanizing

This is the coating of low carbon steels with a layer of zinc. It is an electroplating process as described earlier. The metal deposited is zinc and the process is usually limited to flat and corrugated sheets. It is quicker and cheaper than hot-dip galvanizing, but the coating is thinner.

Any corrosive attack on galvanized products eats away the zinc in preference to the iron. The zinc is said to be sacrificial. To prolong the life of galvanized sheeting it should be painted to protect the zinc itself from the atmosphere.

Surface coatings and finishes

A surface finish can be applied to many engineered products in order to improve its appearance, durability or corrosion resistance. We shall look at a variety of different finishes and how they are applied.

Grinding

A grinding wheel consists of abrasive particles bonded together. It does not 'rub' the metal away, it cuts the metal like any other cutting tool. Each abrasive particle is a cutting tooth. Imagine an abrasive wheel to be a milling cutter with thousands of teeth. Wheels are made in a variety of shapes and sizes. They are also available with a variety of abrasive particle materials and a variety of bonds. It is essential to choose the correct wheel for any given job.

Figure 3.103(a) shows two types of electrically powered portable grinding machines used for dressing welds and for fettling castings. Care must be taken in its use and the operator should wear a suitable grade of protective goggles and a filter type respirator.

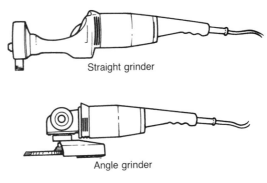

Straight grinder

Angle grinder

(a) Portable grinding machines

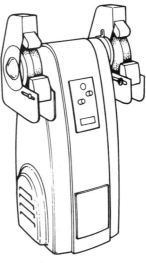

(b) Off-hand grinding machine

Figure 3.103 *Grinding machines*

Figure 3.103(b) shows a double-ended, off-hand tool grinder. This is used for sharpening drills and lathe tools and other small tools and marking out instruments. It is essential to check that the guard is in place and that the visor and tool rest are correctly adjusted before commencing to use the machine. Grinding wheels can only be changed by a trained and certificated person.

Polishing

Polishing produces an even better finish than grinding but only removes the minutest amounts of metal. It only produces a smooth and shiny surface finish, the geometry of the surface is uncontrolled. Polishing is used to produce decorative finishes, to improve fluid flow through the manifolds of racing engines, and to remove machining marks from surfaces that cannot be precision ground. This is done to reduce the risk of fatigue failure in highly stressed components.

Figure 3.104 shows a typical polishing lathe. It consists of an electric motor with a double ended extended spindle. At each end of the spindle is a tapered thread onto which you screw the polishing mops. The mops may be made up from discs of leather (basils) or discs of cloth (calico mops). Polishing compound in the form of 'sticks' is pressed against the mops to impregnate them with the abrasive.

The components to be polished are held against the rapidly spinning mops by hand. Because the mops are soft and flexible they will follow the contours of complex shaped components. It is essential that dust extraction equipment is fitted to the polishing lathe and that the operator wears eye protection and a filter type dust mask. The process should only be carried out by a skilled polisher or under close supervision.

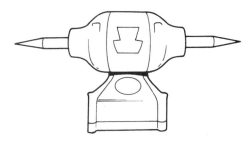

Figure 3.104 *A typical polishing lathe*

Coating

Electroplating has already been discussed and is the coating of metal components with another metal that is more decorative and/or corrosion resistant. Hot-dip galvanizing coats low carbon steels with zinc without using an electroplating process.

Hot-dip galvanizing

Hot-dip galvanizing is the original process used for zinc coating buckets, animal feeding troughs and other farming accessories. It is also used for galvanized sheeting. The work to be coated is chemically cleaned, fluxed and dipped into the molten zinc. This forms a coating on the work. A small percentage of aluminium is added to the zinc to give the traditional bright finish. The molten zinc also seals any cut edges and joints in the work and renders them fluid tight. Metal components may also be coated with non-metallic surfaces.

Oxidizing

Oil blueing

Steel components have a natural oxide film due to their reaction with atmospheric oxygen. This film can be thickened and enhanced by heating the steel component until it takes on a dark blue colour. Then immediately dip the component into oil to seal the oxide film. This process does not work if there is any residual mill scale on the metal surfaces.

Chemical blacking

Alternatively, an even more corrosion resistant oxide film can be applied to steel components by chemical blacking. The components are cleaned and degreased. They are then immersed in the oxidizing chemical solution until the required film thickness has been achieved. Finally the treated components are rinsed, dewatered and oiled. Again, the process only works on bright surfaces.

Plastic coating

Plastic coatings can be both functional, corrosion resistant and decorative. The wide range of plastic materials available in a wide variety of colours and finishes provides a designer with the means of achieving one or more of the following:

- abrasion resistance
- cushioning effects with coatings up to 6 mm thick
- electrical and thermal insulation
- flexibility over a wide range of temperatures
- non-stick properties (Teflon PTFE coatings)
- permanent protection against weathering and atmospheric pollution, resulting in reduced maintenance costs
- resistance to corrosion by a wide range of chemicals
- the covering of welds and the sealing of porous castings.

To ensure success, the surfaces of the work to be plasticized must be physically and chemically clean and free from oils and greases. The surfaces to be plasticized must not have been plated, galvanized or oxide treated.

Fluidized bed dipping

Finely powdered plastic particles are suspended in a current of air in a fluidizing bath as shown in Figure 3.105. The powder continually bubbles up and falls back and looks as though it is boiling. It offers

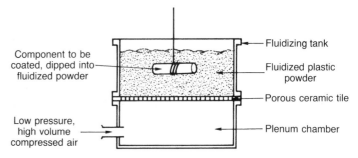

Figure 3.105 *Fluidized bed dipping*

no resistance to the work to be immersed in it. The work is preheated and immersed in the powder. A layer of powder melts onto the surface of the metal to form a homogeneous layer.

Liquid plastisol dipping

This process is limited to PVC coating. A plastisol is a resin powder suspended in a plastisol and no dangerous solvents are present. The preheated work is suspended in the PVC plastisol until the required thickness of coating has adhered to the metal surface.

Painting

Painting is used to provide a decorative and corrosion resistant coating for metal surfaces. It is the easiest and cheapest means of coating that can be applied with any degree of permanence. A paint consists of three components:

- *Pigment* The finely powered pigment provides the paint with its opacity and colour.
- *Vehicle* This is a film-forming liquid or binder in a volatile solvent. This binder is a natural or synthetic resinous material. When dry (set) it must be flexible, adhere strongly to the surface being painted, corrosion resistant and durable.
- *Solvent (thinner)* This controls the consistency of the paint and its application. It forms no part of the final paint film as it totally evaporates. As it evaporates it increases the concentration of catalyst in the 'vehicle' causing it to change chemically and set.

A paint system consists of the following components:

- *Primer* This is designed to adhere strongly to the surface being painted and to provide a key for the subsequent coats. It also contains anti-corrosion compounds.
- *Putty or filler* This is used mainly on castings to fill up and repair blemishes. It provides a smooth surface for subsequent paint coats.
- *Undercoat* This builds up the thickness of the paint film. To produce a smooth finish, more than one undercoat is used with careful rubbing down between each coat. It also gives richness and opacity to the colour.
- *Top coat* This coat is decorative and abrasion resistant. It also seals the paint film with a waterproof membrane. Modern top coats are usually based on acrylic resins or polyurethane rubbers.

There are four main groups of paint:

- *Group 1* The vehicle is polymerized (see thermosetting plastics) into a solid film by reaction with atmospheric oxygen. Paints that set naturally in this way include traditional linseed oil based paints, oleo-resinous paints, and modern general purpose air drying paints based on modified alkyd resins.
- *Group 2* This group of paints is based on amino-alkyd resins that do not set at room temperatures but they have to be *stoved* at 110–150°C. When set these paints are much tougher than group 1 air-drying paints.
- *Group 3* These are the 'two-pack' paints. Polymerization starts to occur as soon as the catalyst is mixed with the paint immediately before use. Modern 'one-pack' versions of these paints have the catalyst diluted with a volatile solvent as mentioned earlier. The solvent evaporates after the paint has been spread and, when the concentration of the catalyst reaches a critical level, polymerization takes place and the paint sets.
- *Group 4* These paints dry by evaporation of the solvent and no polymerization occurs. An example is the cellulose paint used widely at one time for spray painting motor car body panels. Lacquers also belong to this group but differ from all other paints in that dyes are used as the colorant instead of pigments.

Paints may be applied by brushing, spraying or dipping. Whatever method is used, great care must be taken to ensure adequate ventilation. Not only can the solvents produce narcotic effects, but inhaling dried particles and liquid droplets of paint can cause serious respiratory diseases. The appropriate protective clothing, goggles and face masks must be used. Paints are also highly flammable and the local fire-prevention officer must be consulted over their storage and use. On no account can smoking be tolerated anywhere near the storage or use of paints.

Test your knowledge 3.70

State the essential difference between grinding and polishing.

Test your knowledge 3.71

State a suitable coating medium and describe its application process for each of the following engineered components:
(a) painting refrigerator body panels
(b) painting pressed steel angle brackets
(c) plastic cladding metal tubing for a bathroom towel rail.

Activity 3.15

A manufacturer of model boat kits has asked you to produce a single page instruction sheet on 'How to paint finish your kit'. Produce a word processed instruction sheet and illustrate it using appropriate sketches and drawings. Include a section headed 'Safety'.

Review questions

1 Explain, in simple terms, each of the following properties of materials:
(a) strength
(b) toughness
(c) elasticity
(d) hardness
(e) rigidity.

2   Distinguish between the terms *ductility* and *malleability* when applied to engineering materials.

3   Sketch graphs to show how tensile strength and ductility varies with carbon content for a plain carbon steel. Label your graph.

4   Give examples of suitable non-ferrous metals (with reasons for your choice) for each of the following applications:
(a) an instrument case used to house a portable electronic test set
(b) a water pump to be used with a marine engine
(c) a bus-bar to carry electric current in a steelworks
(d) a screw terminal used in an electric light fitting.

5   Distinguish between the terms *thermosetting plastic* and *thermoplastic* when applied to engineering materials.

6   Give examples of suitable plastic materials (with reasons for your choice) for each of the following applications:
(a) a car tow rope
(b) an engine cover for use on a light aircraft
(c) an electrical junction box
(d) the body of a 'ride-on' toy.

7   Sketch the construction of a riveted lap joint.

8   Explain what is meant by a *residual current detector (RCD)*.

9   Sketch a micrometer caliper. Label your drawing showing each of the main features.

10   Explain, with the aid of a diagram, what is meant by the term *rake angle* when applied to a cutting tool.

11   Sketch a pillar drill. Label your drawing showing each of the main features.

12   Describe THREE operations that can be performed using a centre lathe.

13   Explain what is meant by the term *soft soldering*. Why is this process different from *fusion welding*?

14   Describe THREE different methods of making electrical joints.

15   Explain what is meant by the terms:
(a) hardening
(b) tempering.

16   Describe, briefly, the main stages in the production of a printed circuit board.

17   List and briefly explain the function of the components used in a paint system.

Unit 4 Applied science and mathematics for engineering

Summary

This unit looks at the basic mathematics and science used by engineers. Whilst working through this unit you will develop your skills in both mathematics and science by solving problems and investigating some basic engineering systems. It is important to note that we have not attempted to separate the unit into sections on mathematics and science. Instead, we have introduced relevant mathematical techniques wherever they are applicable and useful. This unit underpins all other units and it begins by introducing some fundamental concepts that we shall build on later.

Fundamental concepts

SI units

The system of units used in engineering and science, the 'SI System', based on metric units. The SI system was introduced in 1960 and is now adopted by the majority of countries as the official system of measurement. The basic units in the SI system are listed overleaf with their symbols in Table 4.1.

In some applications the basic SI units may be very large or very small. To help make numbers more manageable we use a common set of *prefixes* that denote multiplication or division of the basic unit by a particular amount. Some of the most common multiples are shown in Table 4.2.

Table 4.1 *The basic SI units*

Quantity	Unit (unit symbol)
length	metre (m)
mass	kilogram (kg)
time	second (s)
electric current	ampere (A)
temperature	Kelvin (K)
luminous intensity	candela (cd)
amount of substance	mole (mol)

Table 4.2 *Multiples and sub-multiples*

Prefix	Name	Meaning
G	giga	multiply by 1 000 000 000 (i.e. $\times 10^9$)
M	mega	multiply by 1 000 000 (i.e. $\times 10^6$)
k	kilo	multiply by 1 000 (i.e. $\times 10^3$)
none	—	multiply by 1
m	milli	divide by 1 000 (i.e. $\times 10^{-3}$)
μ	micro	divide by 1 000 000 (i.e. $\times 10^{-6}$)

Index notation

It is often convenient to write multiples of 10 (such as 1 000 or 1 000 000) using indices. This saves writing a large number of noughts. For example 1 000 is the same as $(10 \times 10 \times 10)$ or 10^3. The *index* (3) simply tells us that the number is equal to 10 (the *base*) multiplied by itself three times. As another example 10^6 is the same as $(10 \times 10 \times 10 \times 10 \times 10 \times 10)$ or 1 000 000.

Standard form

A number written with one digit to the left of the decimal point and multiplied by 10 raised to some power is said to be written in *standard form*. Thus: 5837 is written as 5.837×10^3 in standard form, and 0.0045 is written as 4.5×10^{-3} in standard form.

When a number is written in standard form, the first factor is called the *mantissa* and the second factor is called the *exponent*. Thus the number 6.7×10^3 has a mantissa of 6.7 and an exponent of 10^3.

Numbers having the same exponent can be added or subtracted in standard form by adding or subtracting the mantissae and keeping the exponent the same. For example:

$$(1.2 \times 10^4) + (3.4 \times 10^4) = (1.2 + 3.4) \times 10^4 = 4.6 \times 10^4$$

When multiplying or dividing numbers in standard form the mantissae must be multiplied or divided whilst the exponents must be added or subtracted. For example,

$$(2.5 \times 10^3) \times (5 \times 10^2) = (2.5 \times 5) \times (10^{3+2})$$
$$= 12.5 \times 10^5 \text{ or } 1.25 \times 10^6$$

Similarly,

$$\frac{6 \times 10^6}{1.5 \times 10^2} = \frac{6}{1.5} \times 10^{4-2} = 4 \times 10^2$$

Test your knowledge 4.1

Express the following in standard form:
(a) 9871
(b) 0.334
(c) 145760
(d) 0.0005

Length, area, volume and mass

Length is the distance between two points. The standard unit of length is the *metre* (m) although the *centimetre* (cm), *millimetre* (mm) and *kilometre* (km) are often used.

1 cm = 10 mm; 1 m = 100 cm = 1 000 mm; 1 km = 1 000 m

Area is a measure of the size or extent of a plane surface and is measured by multiplying a length by a length. If the lengths are in metres then the unit of area is the square metre, m^2.

$1\ m^2 = 1\ m \times 1\ m = 100\ cm \times 100\ cm = 10\ 000\ cm^2$ or $10^4\ cm^2$

$1\ m^2 = 1\ 000\ mm \times 1\ 000\ mm = 1\ 000\ 000\ mm^2$ or $10^6\ mm^2$

Conversely $1\ cm^2 = 10^{-4}\ m^2$ and $1\ mm^2 = 10^{-6}\ m^2$

Volume is a measure of the space occupied by a solid and is measured by multiplying a length by a length by a length. If the lengths are in metres then the unit of volume is cubic metres, m^3.

$1\ m^3 = 1\ m \times 1\ m \times 1\ m$

$= 100\ cm \times 100\ cm \times 100\ cm = 10^6\ cm^3$

$= 1\ 000\ mm \times 1\ 000\ mm \times 1\ 000\ mm$

$= 10^9\ mm^3$

Conversely, $1\ cm^3 = 10^{-6}\ m^3$ and $1\ mm^3 = 10^{-9}\ m^3$

Another unit used to measure volume, particularly with fluids, is the litre, l, where 1 l = 1000 cm^3.

Mass is the amount of matter in a body and is measured in *kilograms*, kg.

1 kg = 1 000 g   (or conversely, 1 g = 10^{-3} kg)

and

1 tonne (t) = 1 000 kg

Example 4.1

Express (a) a length of 0.055 metres in mm, and (b) a length of 12 mm in metres.

(a) 1 mm = 10^{-3} m or 1 m = 10^3 mm

Hence 0.055 m = 0.055 x 10^3 mm = 0.055 x 1000 = 55 mm

Note that here we have multiplied the length expressed in m by 1000 to find the corresponding value in mm.

(b) 1 m = 10^3 mm or 1 mm = 10^{-3}m

Hence 12 mm = 12 x 10^{-3} m = $\dfrac{12}{10^3}$ m = $\dfrac{12}{1000}$ m = 0.012 m

Note that we have divided the length expressed in mm by 1000 to find the corresponding value in m.

Test your knowledge 4.2

Determine the area of a steel plate 0.4 m long by 0.15 m wide. Express your answer in:

(a) m^2
(b) cm^2
(c) mm^2

Test your knowledge 4.3

A tank contains 1.6 litres of fuel. Determine the volume of fuel and express your answer in:

(a) cm^3
(b) m^3
(c) mm^3

Example 4.2

A cube has sides each of length 0.5 m. Determine the volume of the cube in cubic cm.

Volume of cube = 0.5 m x 0.5 m x 0.5 m = 0.125 m^3

  1 m^3 = 10^6 cm^3,   thus volume = 0.125 x 10^6 cm^3

                                      = 125 000 cm^3

Alternatively,

Volume of cube = 50 cm x 50 cm x 50 cm = 125 000 cm^3

Density

Density is the mass per unit volume of a substance. The symbol used for density is ρ (Greek letter rho) and its units are kg/m^3.

$$\text{Density} = \frac{\text{mass}}{\text{volume}}$$

Formulae

We can replace density, mass and volume with symbols that represent these quantities. For example, we can say that:

$$\rho = \frac{m}{V}$$

where ρ (or 'rho') is the density in kg/m^3, m is the mass in kg and V is the volume in m^3.

We can check that the formula is correct by writing it in terms of the units:

$$(\text{kg/m}^3) = (\text{kg}) / (\text{m}^3)$$

You should note that, when written in terms of the units, the left hand side of the equation is the same as the right hand side of the equation. This is a useful technique because it can help you to confirm that a formula you are not sure about is actually correct!

Within the formula that we have just met, ρ, m and V are often referred to as *variables*. If we need to find m or V we can simply rearrange the formula to make the *unknown variable* the subject of the formula. This can be done quite easily. For example, let's make m the subject of the formula:

$$\rho = \frac{m}{V}$$

We won't unbalance the formula as long as we do the same thing on each side of the '=' sign. If we multiply *each side* by V we will arrive at:

$$\rho \times V = \frac{m}{V} \times V$$

The V terms on the right hand side will cancel, leaving us with:

$$\rho \times V = m \times \frac{V}{V} = m \times 1 = m$$

Thus $m = \rho \times V$

If we now divide *each side* by ρ we will arrive at:

$$\frac{m}{\rho} = \frac{\rho \times V}{\rho} = \frac{\rho}{\rho} \times V = 1 \times V = V$$

Thus $V = \dfrac{m}{\rho}$

Being able to rearrange formulae is very important, as we shall see later!

Rearranging formulae

The rules for rearranging formulae are quite simple. You can:

- multiply both sides of a formula by the same amount
- divide each side of a formula by the same amount
- add the same amount to each side of a formula
- subtract the same amount from each side of a formula
- square each side of a formula
- square root each side of a formula.

The important thing to remember is that you must treat both sides of the formula exactly the same. Otherwise you will 'unbalance' the formula and it will no longer hold true.

You can also exchange the terms on each side of the '=' sign. In other words:

$$A \times B = C \quad \text{is the same as} \quad C = A \times B$$

Evaluating formulae

In engineering you will use formulae to help you understand what is happening and to allow you to predict what might happen under a certain set of conditions. In either case you will have to substitute given values for known variables in order to find the value of an unknown variable (it is usually convenient for this variable to have first been made the subject of the formula). The following examples show you how this is done:

Example 4.3

Determine the density of a material if 100 cm³ of it has a mass is 0.2 kg.

Volume = 100 cm³ = 100 × 10⁻⁶ m³; mass = 0.2 kg

$$\text{Density} = \frac{\text{mass}}{\text{volume}} = \frac{0.2}{100 \times 10^{-6}} = \frac{0.2}{100} \times 10^6 \text{ kg/m}^3$$

Thus density = 0.002 × 10⁶ kg/m³ = **2000 kg/m³**

Note that dividing a number by 10⁻⁶ is the same as multiplying it by 10⁶.

Test your knowledge 4.4

1 Determine the density of 80 cm³ of cast iron if its mass is 0.5 kg.

2 Determine the volume, in litres, of 35 kg of petrol of density 700 kg/m³.

3 A piece of metal 0.5 m long, 0.1 m wide and 0.01 m thick has a mass of 2 kg. What is the density of the metal?

Example 4.4

The density of aluminium is 2700 kg/m³. Calculate the mass of a block of aluminium if it has a volume of 0.01 m³.

Density, ρ = 2700 kg/m³; volume V = 0.01 m³

Since density = mass/volume, then mass = density × volume.

Hence $m = \rho V$ = 2700 kg/m³ × 0.01 m³ = **27 kg**

Electrical science

Charge and current

All *atoms* consist of *protons*, *neutrons* and *electrons*. The protons, which have positive electrical charges, and the neutrons, which have no electrical charge, are contained within the *nucleus*. Removed from the nucleus are minute negatively charged particles called electrons. Atoms of different materials differ from one another by having different numbers of protons, neutrons and electrons. An equal number of protons and electrons exists within an atom and it is said to be electrically balanced, as the positive and negative charges cancel each other out. When there are more than two electrons in an atom the electrons are arranged into *shells* at various distances from the nucleus.

All atoms are bound together by powerful forces of attraction existing between the nucleus and its electrons. Electrons in the outer shell of an atom, however, are attracted to their nucleus less powerfully than are electrons whose shells are nearer the nucleus.

It is possible for an atom to lose an electron; the atom, which is now called an *ion*, is not now electrically balanced, but is positively charged and is thus able to attract an electron to itself from another atom. Electrons that move from one atom to another are called free electrons and such random motion can continue indefinitely. However, if an electric pressure or *voltage* is applied across any material there is a tendency for electrons to move in a particular direction.

This movement of free electrons, known as *drift*, constitutes an electric current flow. Thus *current is the rate of movement of charge.*

Conductors are materials that have electrons that are loosely connected to the nucleus and can easily move through the material from one atom to another. *Insulators* are materials whose electrons are held firmly to their nucleus.

The unit used to measure the *quantity of electrical charge Q* is called the *coulomb* (C) (where 1 coulomb = 6.24×10^{18} electrons). If the drift of electrons in a conductor takes place at the rate of one coulomb per second the resulting current is said to be a current of one ampere. Thus,

   1 ampere = 1 coulomb per second

or

   1 A = 1 C/s

Hence

   1 coulomb = 1 ampere second   or   1 C = 1 A s

Generally, if I is the current in amperes and t the time in seconds during which the current flows, then $I \times t$ represents the quantity of electrical charge in coulombs, i.e. quantity of electrical charge transferred.

   $Q = I \times t$ coulombs

For a continuous current to flow between two points in a circuit a *potential difference* (p.d.) or *voltage* (V) is required between them; a complete conducting path is necessary to and from the source of electrical energy. The unit of p.d. is the volt (V). Figure 4.1 shows a cell connected across a filament lamp. The cell provides an *electromotive force* (e.m.f.) that 'drives' the current around the circuit. The voltmeter (V) indicates the potential difference across the lamp whilst the ammeter (A) indicates the current flowing. Current flow, by convention, is considered as flowing from the positive terminal of the cell around the circuit to the negative terminal.

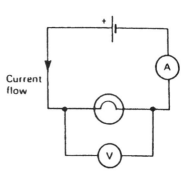

Figure 4.1

The flow of electric current is subject to friction. This friction, or opposition, is called *resistance* (R) and is the property of a conductor that limits current. The unit of resistance is the *ohm* (Ω). 1 ohm is defined as the resistance that will have a current of 1 ampere flowing through it when 1 volt is connected across it, i.e.

$$\text{resistance, } R = \frac{\text{potential difference, } V}{\text{current, } I}$$

Potential difference and resistance

Ohm's law states that the current I flowing in a circuit is directly proportional to the applied voltage V and inversely proportional to the resistance R, provided the temperature remains constant. Thus,

$$I = \frac{V}{R} \quad \text{or} \quad V = IR \quad \text{or} \quad R = \frac{V}{I}$$

Example 4.5

The current flowing through a resistor is 0.8 A when a voltage of 20 V is applied. Determine the value of the resistance.

From Ohm's law,

$$\text{resistance } R = \frac{V}{I} = \frac{20}{0.8} = \mathbf{25\ \Omega}$$

Example 4.6

Determine the potential difference which must be applied to a 20 Ω resistor in order that a current of 10 mA may flow.

Resistance $R = 20\ \Omega$

Current $I = 10\ \text{mA} = 10 \times 10^{-3}\ \text{A} = 0.01\ \text{A}$

From Ohm's law, potential difference, $V = IR = (0.01)(20) = \textbf{0.2 V}$

Example 4.7

A coil has a current of 50 mA flowing through it when the applied voltage is 12 V. What is the resistance of the coil?

$$\text{Resistance, } R = \frac{V}{I} = \frac{12}{50 \times 10^{-3}} = \textbf{240 }\Omega$$

Example 4.8

An electric kettle has a resistance of 30 Ω. What current will flow when it is connected to a 240 V supply?

$$\text{Current, } I = \frac{V}{R} = \frac{220}{55} = \textbf{4 A}$$

Example 4.9

A current of 10 μA flows in a circuit having a resistance of 4.3 MΩ. What voltage drop appears across the resistance?

Voltage, $V = IR = (10 \times 10^{-6}) \times (4.3 \times 10^{6}) = 43\ \text{V}$

or $V = (10 \times 4.3) \times (10^{-6} \times 10^{6}) = 43 \times 10^{-6+6} = 43 \times 10^{0} = 43 \times 1\ \text{V}$

Hence $V = \textbf{43 V}$

Test your knowledge 4.6

1 A 6 V battery is connected across a resistor and causes a current of 1.5 mA to flow. Determine the resistance of the resistor. If the voltage is now reduced to 4 V, what will be the new value of the current flowing?

2 What is the resistance of a coil which draws a current of:
(a) 0.5 A and (b) 1.2 mA from a 24 V supply?

Straight line graphs

We often use graphs to plot experimental data. The shape of the graph helps us to understand how the two quantities are related. Where two quantities are directly proportional to one another the graph plotted to show their relationship will take the form of a straight line. Putting this the other way round, if we plot a graph and it takes the form of a straight line we can infer that the two plotted quantities are directly proportional to one another. If this is beginning to sound a little complex the following example may help.

Example 4.10

The current/voltage relationship for a resistor is shown in Figure 4.2. Determine the resistance of the resistor.

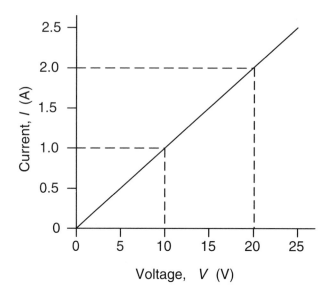

Figure 4.2

We can see that there is a straight line relationship between the current, *I*, and the voltage, *V*. The resistance, *R*, can be found by taking any pair of corresponding values for *I* and *V*.

Taking 1 A and 10 V as a corresponding pair gives:

$$R = \frac{V}{I} = \frac{10\,V}{1\,A} = 10\,\Omega$$

Similarly, taking 2 A and 20 V as a corresponding pair gives:

$$R = \frac{V}{I} = \frac{20\,V}{2\,A} = 10\,\Omega$$

Notice that the result is the same!

The steepness (or *slope*) of the *I* / *V* graph will depend on the value of resistance in the circuit. The steeper the slope the *lower* the resistance will be, and vice versa. We can then say that the resistance, *R*, is *inversely proportional* to the slope of the graph. Hence:

$$\text{Resistance, } R \propto \frac{1}{\text{slope}}$$

where ∝ is a symbol that indicates 'is proportional to'.

Example 4.11

The current/voltage relationship for two resistors A and B is as shown in Figure 4.3. Determine the value of the resistance of each resistor.

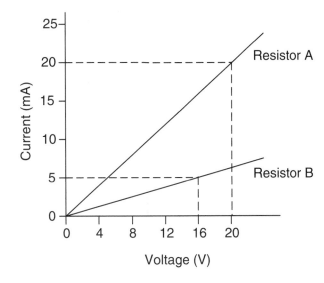

Figure 4.3

For resistor A,

$$R = \frac{V}{I} = \frac{20\ V}{20\ mA} = \frac{20\ V}{0.02\ A} = 1000\ \Omega\ \text{or}\ \textbf{1 k}\boldsymbol{\Omega}$$

For resistor B,

$$R = \frac{V}{I} = \frac{16\ V}{5\ mA} = \frac{16\ V}{0.005\ A} = 3200\ \Omega\ \text{or}\ \textbf{3.02 k}\boldsymbol{\Omega}$$

Note how the lower resistance value has the steeper slope.

Engineers need to be able to plot graphs from experimental data. It is useful to start with a table of corresponding pairs of values. Then mark your axes on the graph by drawing two lines at right angles and marking values at suitable intervals along each of the axes that you have constructed.

Next, take each pair of values, locate them on the axes of the graph and then plot their points of intersection. The best way of doing this is to use a cross using a hard pencil. Finally, join the points of intersection with a line which can be hand drawn (in the case of a curve) or using a rule (in the case of a straight line). In either case, you may need to take the line of 'best fit'.

Finally, make sure that you label your axes clearly! The next example shows how this is done.

Example 4.12

The following results were obtained during an experiment on a resistor:

Voltage (V)	0	10	20	30	40	50
Current (A)	0	0.082	0.164	0.246	0.328	0.410

Plot a graph to show how the current (I) varies with the voltage (V) and confirm that the relationship is linear. Also find the value of the resistance.

First we need to plot the graph as shown below:

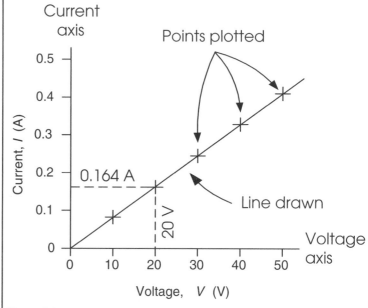

Figure 4.4

Having plotted the graph we can now confirm that the relationship is a linear one (current is directly proportional to voltage).

Finally, we can determine the resistance value by taking a pair of current and voltage values and substitute them into the formula, as follows:

When $V = 50$ V, $I = 0.410$ A.

Hence, $R = \dfrac{V}{I} = \dfrac{50 \text{ V}}{0.410 \text{ A}} = \textbf{122 } \boldsymbol{\Omega}$

Later in this unit we shall investigate graphs that are 'non-linear'. For now you need only remember that a straight line graph corresponds to a linear relationship between the two plotted quantities.

Series circuits

Figure 4.5 shows three resistors, R_1, R_2 and R_3 connected end to end, i.e. in *series*, with a battery source of V volts. Since the circuit is closed a current I will flow and the voltage across each resistor may be determined from the voltmeter readings V_1, V_2 and V_3.

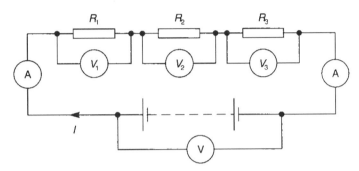

Figure 4.5

In a series circuit:

(a) the current I is the same in all parts of the circuit and hence the same reading is found on each of the ammeters shown, and

(b) the sum of the voltages V_1, V_2 and V_3 is equal to the total applied voltage, V, i.e.

$$V = V_1 + V_2 + V_3$$

From Ohm's law:

$$V_1 = IR_1, \; V_2 = IR_2, \; V_3 = IR_3 \text{ and } V = IR$$

where R is the total circuit resistance.

Since $V = V_1 + V_2 + V_3$, then $IR = IR_1 + IR_2 + IR_3$. Dividing throughout by I gives

$$R = R_1 + R_2 + R_3$$

Thus for a series circuit, the total resistance is obtained by adding together the values of the separate resistances.

Test your knowledge 4.7

A 12 V battery is connected to a circuit comprising three series-connected resistors having resistance of 14 Ω, 16 Ω and 18 Ω. Determine the current supplied to the circuit. Also determine the voltage dropped across the 16 Ω resistor.

Example 4.13

Resistors of 10 Ω, 15Ω and 25 Ω are connected in series across a supply of 25 V. Determine the total resistance of the circuit and the current supplied.

Total resistance $R = R_1 + R_2 + R_3 = 10 + 15 + 25 = 50 \text{ Ω}$

Current flowing $I = \dfrac{V}{R} = \dfrac{25}{50} = 0.5 \text{ A}$

Example 4.14

For the circuit shown in Figure 4.6, determine (a) the battery voltage V, (b) the total resistance of the circuit, and (c) the values of resistance of resistors R_1, R_2 and R_3, given that the p.d.s across R_1, R_2 and R_3 are 5 V, 2 V and 6 V, respectively.

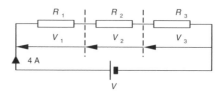

Figure 4.6

(a) Battery voltage $V = V_1 + V_2 + V_3 = 5 + 2 + 6 =$ **13 V**

(b) Total circuit resistance $R = \dfrac{V}{I} = \dfrac{13\ V}{4\ A} =$ **3.25 Ω**

(c) Resistance $R_1 = \dfrac{V_1}{I} = \dfrac{5\ V}{4\ A} =$ **1.25 Ω**

   Resistance $R_2 = \dfrac{V_2}{I} = \dfrac{2\ V}{4\ A} =$ **0.5 Ω**

   Resistance $R_3 = \dfrac{V_3}{I} = \dfrac{6\ V}{4\ A} =$ **1.5 Ω**

   (Check: $R_1 + R_2 + R_3 = 1.25 + 0.5 + 1.5 = 3.25\ Ω = R$)

Activity 4.1

A manufacturer of Christmas decorations has asked you to advise on the development of a new set of Christmas tree lights. The lights are to be connected in series in order to operate from a 240 V mains supply and the company has been able to locate a supply of miniature light bulbs rated at 12 V, 0.1 A. The company has asked you to provide answers to the following questions:

(a) How many light bulbs will be needed in each set of Christmas tree lights?
(b) What current will be taken from the mains supply from a set of lights?
(c) What is the resistance of an individual light bulb when operating?
(d) What is the total resistance of a set of Christmas tree lights when operating?

Present your results in the form of worked calculations with solutions.

Parallel circuits

Figure 4.7 shows three resistors, R_1, R_2 and R_3 connected across each other, i.e. in parallel, across a battery source of V volts.

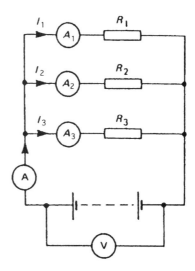

Figure 4.7

In a parallel circuit:

(a) the sum of the currents I_1, I_2 and I_3 is equal to the total circuit current, I, i.e.

$$I = I_1 + I_2 + I_3$$

(b) the source p.d., V volts, is the same across each of the resistors.

From Ohm's law:

$$I_1 = V/R_1 , \quad I_2 = V/R_2 , \quad I_3 = V/R_3 \quad \text{and} \quad I = V/R$$

where R is the total circuit resistance. Since

$$I = I_1 + I_2 + I_3$$

then

$$V/R = V/R_1 + V/R_2 + V/R_3$$

Dividing throughout by V gives

$$1/R = 1/R_1 + 1/R_2 + 1/R_3$$

This equation must be used when finding the total resistance R of a parallel circuit. For the special case of *two resistors in parallel*:

$$1/R = 1/R_1 + 1/R_2 = \frac{R_1 + R_2}{R_1 R_2}$$

Thus $\quad R = \dfrac{R_1 R_2}{R_1 + R_2} = \dfrac{product}{sum}$

Example 4.15

For the circuit shown in Figure 4.8, determine (a) the reading on the ammeter, and (b) the value of resistor R_2.

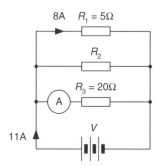

Figure 4.8

The voltage dropped across R_1 is the same as the supply voltage V. Hence supply voltage, $V = 8 \times 5 = 40$ V.

(a) Reading on ammeter

$$I = \frac{V}{R_3} = \frac{40 \text{ V}}{20 \text{ } \Omega} = \textbf{2 A}$$

(b) Current flowing through $R_2 = 11 - 8 - 2 = 1$ A, hence

$$R_2 = \frac{V}{I_2} = \frac{40 \text{ V}}{1 \text{ A}} = \textbf{40 } \boldsymbol{\Omega}$$

Activity 4.2

A light aircraft manufacturer has asked you to help with the design of the aircraft's electrical power system. The power system operates from a 12 V battery and it comprises the following parallel-connected equipment:
(a) an aircraft radio which requires a current of 6 A at 12 V
(b) two navigation lights, each rated at 12 V, 4 A
(c) a cabin light rated at 12 V, 2 A.

The company has asked you to provide answers to the following questions:

(a) How much current must the battery supply?
(b) What is the resistance of each item of electrical equipment?
(c) What is the effective resistance of the complete aircraft electrical system?

Present your results in the form of worked calculations with solutions.

Look at the graph shown in Figure 4.9 and use it to answer the following questions:

(a) Which of the materials show an increase of resistance with temperature?
(b) Which of the materials show a decrease of resistance with temperature?
(c) Which of the materials has a resistance that remains constant regardless of temperature?
(d) Which material shows the greatest change in resistance?

Temperature coefficient

The resistance of metal conductors like copper and brass tends to increase with temperature. The resistance of some other conductors (like carbon) and many insulators (which tend to have a very high resistance) tend to decrease with temperature. This relationship is best illustrated graphically:

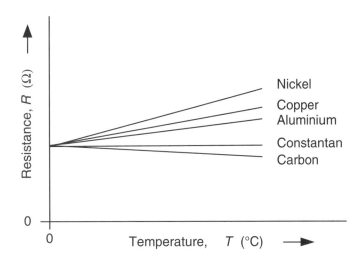

Figure 4.9 *Variation of resistance with temperature for various materials*

We need some way of determining the value of resistance at any given temperature. This is particularly important for applications in which the temperature varies widely. Let's consider one of the most commonly used electrical materials, copper. If we were to carry out an experiment on a copper wire which has a resistance of $1\ \Omega$ at $0°C$ we would find that the resistance increases to $1.43\ \Omega$ at $100°C$, as shown in Figure 4.10.

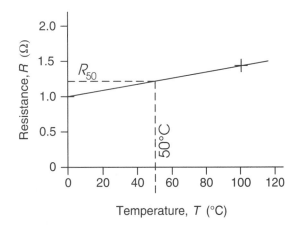

Figure 4.10 *Variation of resistance with temperature for a copper wire having a resistance of $1\ \Omega$ at $0°C$.*

Since the relationship between resistance and temperature is approximately linear over a normal temperature range, we could use the graph in Figure 4.10 to find the resistance at any temperature. Note that:

Resistance at 1°C, $R_0 = 1 \, \Omega$
Resistance at 100°C, $R_{100} = 1.43 \, \Omega$
Resistance at 50°C, $R_{50} = 1.22 \, \Omega$

Using a graph to determine the resistance at any temperature is not particularly convenient. What we really need is a means of quantifying the change in resistance for any particular material and a formula that allows us to determine the resistance at a temperature, T, for any for any given R_0 value. To do this we need to define a constant for a particular material called the *temperature coefficient of resistance*.

The temperature coefficient of resistance of a material is the increase in the resistance of a $1 \, \Omega$ resistor made from that material when it is subjected to a rise of temperature of 1°C. The symbol used for the temperature coefficient of resistance is α (or 'alpha'). Thus, if some copper wire of resistance $1 \, \Omega$ is heated through 1°C and its resistance is then measured as 1.0043°C then α = 0.0043 Ω/Ω/°C for copper. The units are usually expressed only as 'per °C', i.e. α = 0.0043°C for copper. We have already seen how, if a $1 \, \Omega$ resistor of copper is heated through 100°C, its resistance at 100°C would be $1.43 \, \Omega$. Hence,

$$R_{100} = R_0 + (R_0 \times 100 \times 0.0043) = 1 + 0.43 = 1.43 \, \Omega$$

A more general equation, in terms of temperature, T °C, and temperature coefficient, α /°C, would be:

$$R_T = R_0 + (R_0 \times T \times \alpha)$$

This can be rearranged to give:

$$R_T = R_0 + (R_0 \, \alpha \, T)$$

or

$$R_T = R_0(1 + \alpha T)$$

Some values for temperature coefficient are listed in Table 4.3.

Test your knowledge 4.9

(a) A nickel wire has a resistance of 220 Ω at 0°C. Determine its resistance at 80°C.
(b) A copper wire has a resistance of 50 Ω at 100°C. What will its resistance be at 0°C?
(c) A carbon resistor has a resistance of 5 kΩ at 80°C. What will its resistance be at 20°C?

Table 4.3 *Temperature coefficient of resistance for various materials*

Material	α	Material	α
Copper	0.0043/°C	Aluminium	0.0038/°C
Nickel	0.0062/°C	Carbon	-0.00048/°C
Constantan	0/°C	Eureka	0.00001/°C

Example 4.16

The data shown below was obtained during an experiment on a wirewound resistor. Plot a graph showing how the resistance varies with temperature and use it to determine the temperature coefficient of resistance of the material used. Hence determine the value of resistance at a temperature of 125°C.

Temperature (°C)	0	20	40	60	80	100
Resistance (Ω)	1	1.1	1.15	1.22	1.29	1.36

The graph is plotted as shown in Figure 4.11 using a 'line of best fit'.

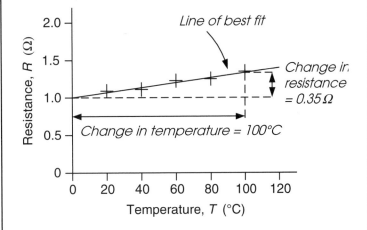

Figure 4.11

We can determine the value of temperature coefficient of resistance, α, from the slope of the graph. To do this we need to construct a triangle and measure the change in resistance that occurs for a given change in temperature. In order to obtain an accurate result it is advisable to use as wide a range of temperature as possible. These measurements show:

Change in resistance = 0.35 Ω (this is actually the same as $R_{100} - R_0$)

Change in temperature = 100°C (this is actually the same as 100 – 0)

Thus the slope of the graph (which in this case is α) is given by:

Slope, α = (0.35 / 1) / 100 = 0.0035 Ω/Ω/°C = 0.0035 /°C

Note that the change in resistance was 0.35 Ω for a starting value (at 0°C of 1Ω). Now to find the value of resistance at a temperature, T, of 125°C:

$$R_T = R_0 (1 + \alpha T) = 1 (1 + 0.0035 \times 125)$$

Hence

$$R_T = 1 (1 + 0.4375) = \mathbf{1.4375 \ \Omega}$$

Look at the graph shown in Figure 4.12 and use it to answer the following questions:

(a) Which of the materials has a large positive value of α?
(b) Which of the materials has a negative value for α?
(c) For which of the materials is α zero?
(d) Which of the materials has a non-linear resistance/ temperature characteristic?
(e) Which THREE materials have positive temperature coefficients?
(f) Which material would be most suited for use as the connecting leads to a high temperature probe? Explain your answer.

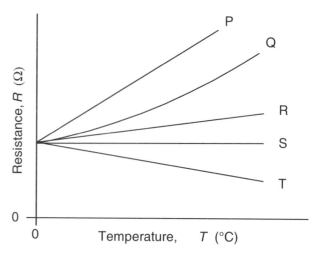

Figure 4.12 *See Test your knowledge 4.10*

Effects of an electric current

The three main effects of an electric current are:

(a) the magnetic effect
(b) the chemical effect
(c) the heating effect.

The first of these three effects, the *magnetic effect*, can be detected by placing a compass close to a wire that is carrying an electric current. The needle of the compass will deflect as it is brought near to the wire showing that there is a magnetic field in the space around the wire. The field set up by the current will continue to exist for as long as the current flows. The magnetic effect has many applications including electric bells, relays, buzzers, lifting magnets, motors, generators, etc.

The second effect, the *chemical effect*, can be demonstrated by placing two copper plate electrodes in a glass jar containing copper sulphate solution. If the electrodes are connected to a battery a current will flow and the electrode connected to the negative terminal (we call this the *cathode*) will gain copper whilst the electrode connected to the positive terminal (we call this the *anode*) will lose copper. Other applications that rely on the chemical effect of an electric current are simple batteries and cells.

The third effect, the *heating effect*, can be demonstrated by connecting a low-value resistor to a battery. The current flowing through the resistor will a cause it to become warm. In this particular arrangement the chemical energy supplied by the battery is being converted into electrical energy and this, in turn, is being converted into heat energy. Applications of the heating effect of an electric current include water heaters, electric cookers, hot plates, electric fires, electric irons, soldering irons, and electric kettles.

Electromagnetism

A permanent magnet is a piece of ferromagnetic material (such as iron, nickel or cobalt) that has the property of attracting other pieces of these materials. The area around a magnet is called the magnetic field and it is in this area that the effects of the magnetic force produced by the magnet can be detected. The magnetic field of a bar magnet can be represented pictorially by the 'lines of force' (or lines of 'magnetic flux') as shown in Figure 4.13. Such a field pattern can be produced by placing iron filings in the vicinity of the magnet or can be plotted using a compass. The field direction is taken as the movement of a 'free' north pole (i.e. from the north pole to the south pole) as indicated by the arrows in Figure 4.13.

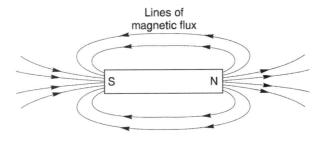

Figure 4.13 *The magnetic field around a bar magnet*

As we have just mentioned, magnetic fields are produced by electric currents as well as by permanent magnets. A straight wire carrying a current will produce a magnetic field pattern that is circular with the current-carrying conductor at the centre. The magnetic field can be concentrated by making several loops of wire into a *solenoid* or *coil* as shown in Figure 4.14. If the solenoid is wound on an ferromagnetic core (iron or steel) an even stronger field is produced. Once again the field can be plotted using iron filings or a compass.

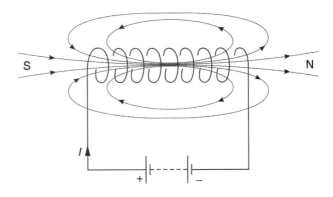

Figure 4.14 *The magnetic field around a solenoid*

Electromagnetic induction

When a conductor is moved across a magnetic field so as to cut through the magnetic filed lines (or *flux*), an electromotive force (e.m.f.) is produced in the conductor (recall that an e.m.f. is what 'drives' current around a circuit). If the conductor forms part of a closed circuit then the e.m.f. produced causes an electric current to flow round the circuit. Hence an e.m.f. (and thus current) is *induced* in the conductor as a result of its movement across the magnetic field. This effect is known as *electromagnetic induction.*

Figure 4.15(a) shows a coil of wire connected to a centre-zero galvanometer, which is a sensitive ammeter with a zero-current position in the centre of the scale. When the magnet is moved at constant speed towards the coil, a deflection is noted on the galvanometer showing that a current has been produced in the coil.

When the magnet is moved at the same speed but away from the coil the same deflection is noted but is in the opposite direction as shown in Figure 4.15(b). When the magnet is held stationary even within the coil no deflection is recorded.

When the magnet is held stationary and the coil is moved towards it, the galvanometer will produce a deflection as shown in Figure 4.15(c). It is also worth noting that:

- When the relative speed is, say, doubled, the galvanometer deflection is doubled.
- When a stronger magnet is used, a greater galvanometer deflection is noted.
- When the number of turns of wire of the coil is increased, a greater galvanometer deflection is noted.

Finally, note that Figure 4.15(c) shows the magnetic field associated with the magnet. As the magnet is moved towards the coil, the magnetic flux of the magnet moves across, or *cuts*, the coil. It is the *relative movement of the magnetic flux and the coil* that causes an e.m.f. (and thus current) to be induced in the coil. This effect is known as *electromagnetic induction.* The laws of electromagnetic induction evolved from experiments such as those described above.

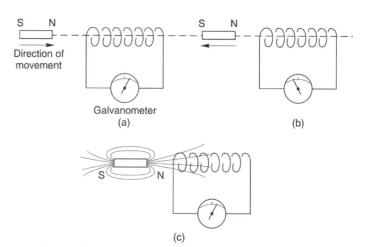

Figure 4.15 *Experiments in electromagnetic induction*

Induced e.m.f.

Faraday's laws of electromagnetic induction state that:

- An induced e.m.f. is set up whenever the magnetic field linking that circuit changes.
- The magnitude of the induced e.m.f. in any circuit is proportional to the rate of change of the magnetic flux linking the circuit.

Lenz's law states that:

> The direction of an induced e.m.f. is always such that it tends to set up a current opposing the motion or the change of flux responsible for inducing that e.m.f.

An alternative method to Lenz's law of determining relative directions is given by Fleming's *R*ight-hand rule (often called the gene*R*ator rule) which states:

> Let the thumb, first finger and second finger of the right hand be extended such that they are all at right angles to each other (as shown in Figure 4.16). If the first finger points in the direction of the magnetic field, the thumb points in the direction of motion of the conductor relative to the magnetic field, then the second finger will point in the direction of the induced e.m.f.

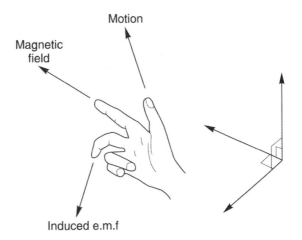

Figure 4.16 *Faraday's right-hand rule*

To help you remember Figure 4.16 you can say that:

- *F*irst finger = *F*ield
- Thu*M*b = *M*otion
- S*E*cond finger = *E*.m.f.

In a *generator*, conductors forming an electric circuit are made to move through a magnetic field. By Faraday's law an e.m.f. is induced in the conductors and thus a source of e.m.f. is created. A generator converts mechanical energy into electrical energy.

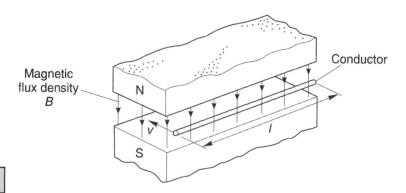

Figure 4.17 *A conductor moving inside a magnetic field*

The induced e.m.f., E, set up between the ends of the conductor shown in Figure 4.17 is given by:

$$E = B\, l\, v \quad \text{volts}$$

where B is the magnetic flux density (i.e. the strength of the magnetic field) measured in teslas, l is the length of conductor in the magnetic field is measured in metres, and v is the conductor velocity is measured in metres per second.

Finally, if the conductor moves at an angle θ° to the magnetic field (instead of at 90° as assumed above) then:

$$E = B\, l\, v \sin \theta$$

Example 4.17

A conductor 300 mm long moves at a uniform speed of 4 m/s at right angles to a uniform magnetic field of flux density 1.25 T. Determine the induced e.m.f. and the current flowing in the conductor when it is connected to a 20 Ω resistance.

Note that, when a conductor moves in a magnetic field it will have an e.m.f. induced in it but this e.m.f. *can only produce a current* if there is a closed path for the current to flow.

(a) Induced e.m.f. $E = B\,l\,v = 1.25\text{ T} \times 0.3\text{ m} \times 4\text{ m/s} = \textbf{1.5 V}$

(b) From Ohm's law

$$I = \frac{V}{R}$$

In this case the voltage dropped across the 20 Ω resistor will be the same as the voltage generated by the conductor moving in the magnetic field. In other words, $V = E$. Thus:

$$I = \frac{V}{R} = \frac{E}{R} = \frac{1.5\text{ V}}{20\ \Omega} = 0.075\text{ A} = \textbf{75 mA}$$

The sine function

The sine of an angle is derived from the ratio of the sides of a right-angled triangle in which the angle appears. Figure 4.18 shows a typical right-angled triangle having sides of length a, b and c. The angle that we are concerned with, θ, has also been marked.

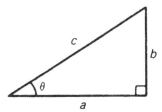

Figure 4.18 *A right-angled triangle*

In Figure 4.18, side a is called the *adjacent* side, side b is called the *opposite* side, and side c is called the *hypotenuse*. Note that, since the triangle is right-angled, we can say that:

$$c^2 = a^2 + b^2$$

The sine of angle θ (or *sin θ*) is defined as:

$$\sin\theta = \frac{\text{opposite}}{\text{hypotenuse}}, \quad \text{i.e. } \sin\theta = \frac{b}{c}$$

By drawing up a table of values of θ from $0°$ to $360°$, a graph of $\sin\theta$ can be plotted. Values obtained with a calculator (correct to three decimal places – which is more than sufficient for plotting graphs), using $30°$ intervals, are shown in the table below:

Table 4.4 *Corresponding values of θ and $\sin\theta$*

θ	$0°$	$30°$	$60°$	$90°$	$120°$	$150°$	$180°$
$\sin\theta$	0	0.500	0.866	1	0.866	0.500	0

θ	$210°$	$240°$	$270°$	$300°$	$330°$	$360°$
$\sin\theta$	−0.500	−0.866	−1	−0.866	−0.500	0

Example 4.18

Use the data in Table 4.4 to plot a graph of $\sin\theta$ for values of θ from $0°$ to $360°$.

The graph is shown in Figure 4.19.

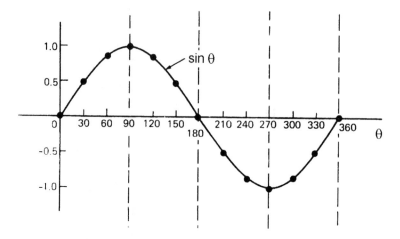

Figure 4.19 *A graph of the sine function*

Direct current and alternating current

Up to this point, the circuits that we have looked at have all derived their supplies from batteries. Batteries consist of cells that convert chemical energy into electrical energy. The voltage and current produced by a battery remain reasonably constant with time (until, of course, the battery eventually goes 'flat' and can no longer supply any charge).

We say that batteries are a source of *direct current* (DC). Direct current is current that flows in one direction only and remains constant with time.

There is another type of current that is commonly used. This is called alternating current (AC). Alternating current flows first in one direction and then in the other, repeating this backwards/forwards cycle as long as it is connected. We can illustrate this by using a graph (this time using time, *t*, as the horizontal axis).

We call graphs of current or voltage plotted against time *waveforms*. Figure 4.20 shows three types of current; DC, AC and DC with AC superimposed. You will find all three types of waveform in most electronic circuits!

The *frequency* of an alternating current is the number of complete cycles (forward/backward) of the current that occur in a time interval of 1 second. Frequency is expressed in hertz (this is the same as 'cycles per second'). The frequency of the AC mains supply in the UK is 50 Hz. This means that 50 forward/backward cycles occur every second.

The period of an alternating current is the time taken for one complete cycle of the current. If 50 cycles of a current occur in one second the time for each cycle must be 1/50 second or 20 ms. Thus we can conclude that:

$$T = 1/f \quad \text{or} \quad f = 1/T$$

Where f is the frequency in hertz and T is the period in seconds.

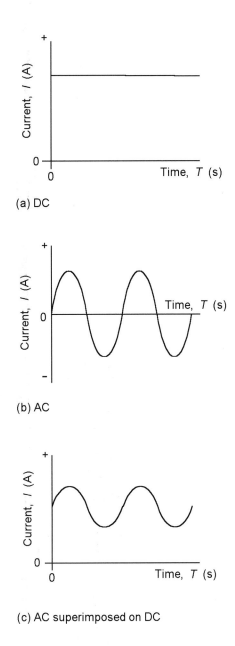

Figure 4.20 *Alternating current (AC) and direct current (DC)*

The *amplitude* (or *peak* value) of an alternating current is the maximum excursion from the mean value. Note that the mean value of a symmetrical AC waveform (such as a sine wave) is zero and thus the amplitude is usually measured from the axis (corresponding to zero current or voltage) as shown in Figure 4.21.

The *peak-peak value* of an alternating current is the difference between the maximum negative value of current and the maximum positive value of current. The peak-peak value is twice the peak value. Thus a waveform with an amplitude (peak value) of 100 V wil have a peak-peak value of 200 V.

Test your knowledge 4.15

Look at the graph shown in Figure 4.22 and use it to answer the following questions:

(a)   What type of current is this?
(b)   What is the amplitude of the current?
(c)   What is the peak-peak value of the current?
(d)   What is the period of the current?
(e)   What is the frequency of the current?

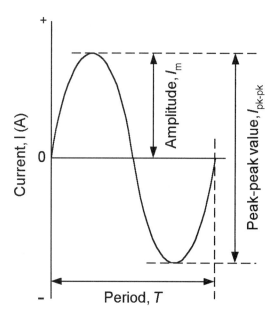

Figure 4.21   *Amplitude, period and peak-peak values*

Test your knowledge 4.16

Sketch the waveforms of:

1.   A sine wave current having a period of 50 ms and an amplitude of 6 A.
2.   A sine wave voltage having a frequency of 1 kHz and a peak-peak value of 120 V.

Label your sketches and include values.

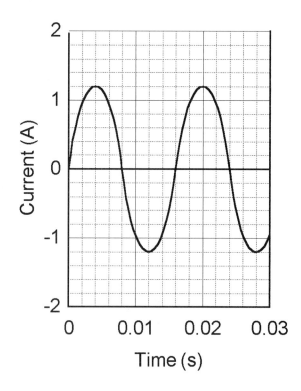

Figure 4.22   *See Test your knowledge 4.15*

The AC generator

Electricity is produced by generators at power stations and then distributed by a vast network of transmission lines (called the National Grid system) to industry and for domestic use. It is easier and cheaper to generate alternating current (AC) than direct current (DC) and AC is more conveniently distributed than DC since its voltage can be readily altered using transformers. Whenever DC is neded in preference to AC, components called rectifiers are used for conversion.

The shape of the AC waveform generated and distributed on the National Grid is a sine wave. Let's explain why this is. Think of a single turn coil that's free to rotate at constant speed between the poles of a magnet system as shown in Figure 4.23. A current is induced in the coil (assuming that the coil is actually connected into a circuit) which varies in magnitude and reverses its direction at regular intervals. The reason for this is shown in Figure 4.24.

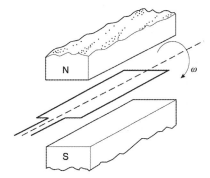

Figure 4.23 *A single loop of wire rotating in a magnetic field*

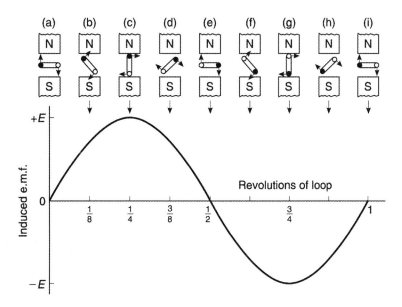

Figure 4.24 *Sinusoidal e.m.f. produced by the loop shown in Figure 4.23*

In positions (a), (e) and (i) the conductors of the loop are effectively moving along the magnetic field, no magnetic flux is cut and hence no e.m.f. is induced. At this point the angle between the direction of motion and the field lines is 0° (recall that sin 0° = 0).

In position (c) maximum e.m.f. is induced. At this point the angle between the direction of motion and the field lines is 90° (recall that sin 90° = 1).

In position (g), maximum flux is cut and hence maximum e.m.f. is again induced. However, using Fleming's right-hand rule, the induced e.m.f. is in the opposite direction to that in position (c) and is thus shown as –E. At this point the angle between the direction of motion and the field lines is 270° (recall that sin 270° = –1).

In positions (b), (d), (f) and (h) some flux is cut and hence some e.m.f. is induced. If all such positions of the coil are considered, in one revolution of the coil, one cycle of alternating e.m.f. is produced as shown. This is the principle of operation of the AC generator or *alternator*.

Finally, it is important to note that the waveform produced by the AC generator has a sine wave shape. This should not be very surprising since you should recall that:

$$E = B \, l \, v \sin \theta$$

from which we can deduce that E is *directly proportional* to sin θ. Hence:

$$E \propto \sin \theta$$

Activity 4.3

Use information sources and catalogues from electronic equipment suppliers and/or kit manufacturers to obtain data on a typical portable metal detector. Write a brief article for a local newspaper explaining the principle on which the metal detector works. Your article should be accompanied by relevant sketches and diagrams and should be suitable for the non-technical reader.

Activity 4.4

Investigate an AC mains adapter that can be used to provide DC power for a typical consumer electronic product. Obtain the specification for the unit in terms of its rated DC output voltage and current. Explain, with the aid of a simple block diagram, how the unit operates and sketch waveforms of the AC input voltage and DC output voltage. Label your waveforms. Present your work in the form of a word processed fact sheet.

Mechanical science

Scalar and vector quantities

Quantities used in engineering and science can be divided into two main groups:

Scalars

Scalar quantities (*scalars*) have a size (or magnitude) only and need no other information to specify them. Thus, 10 centimetres, 50 seconds, 7 litres and 3 kilograms are all examples of scalars.

Vectors

Vector quantities (*vectors*) have both a size or magnitude and a direction, called the line of action of the quantity. Thus, a velocity of 50 kilometres per hour due east, an acceleration of 9.81 metres per second squared vertically downwards and a force of 15 newtons at an angle of 30 degrees are all examples of vectors.

Force

Force is an example of a vector quantity. It thus has both a *magnitude* and a *direction*. A vector can be represented graphically by a line drawn to scale in the direction of the line of action of the force. Vector quantities are often shown by using bold, lower case letters, thus **ab** in Figure 4.25 represents a force of 5 newtons acting in a direction due east. Figure 4.25 is called a *vector diagram*.

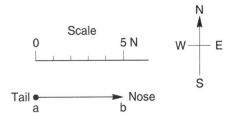

Figure 4.25 *Representing a force*

The resultant of two coplanar forces

When forces are all acting in the same plane, they are called *coplanar*. When forces act at the same time and at the same point, they are called *concurrent forces*. For two forces acting at a point and along the same line, there are two possibilities:

(a) For forces acting in the same direction and having the same line of action, the single force having the same effect as both of the forces, called the *resultant force* or just the *resultant*, is the arithmetic sum of the separate forces. Forces of F_1 and F_2 acting at point P, as shown in Figure 4.26(a), have exactly the same effect on point P as force F shown in Figure 4.26(b), where $F = F_1 + F_2$ and acts in the same direction as F_1 and F_2. Thus F is the resultant of F_1 and F_2.

(b) For forces acting in opposite directions along the same line of action, the resultant force is the arithmetic difference between the two forces. Forces of F_1 and F_2 acting at point P as shown in Figure 4.27(a) have exactly the same effect on point P as force F shown in Figure 4.27(b), where $F = F_2 - F_1$ and acts in the direction of F_2, since F_2 is greater than F_1. Once again, F is the resultant of F_1 and F_2.

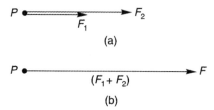

Figure 4.26 *Forces acting on a point in the same direction*

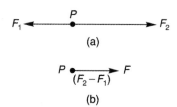

Figure 4.27 *Forces acting on a point in the opposite direction*

Example 4.19

Determine the resultant of two forces of 5 kN and 8 kN acting having the same line of action but (a) acting in the same direction and (b) acting in opposite directions.

The vector diagram is shown in Figure 4.28.

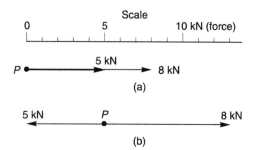

Figure 4.28

Note that in case (a):

   $F = F_1 + F_2 = 5 + 8 = \textbf{13 kN}$ in the direction of the 8 kN force

whereas in case (b):

   $F = F_2 - F_1 = 8 - 5 = \textbf{3 kN}$ in the direction of the 8 kN force.

Vector addition

When two forces do not have the same line of action, the magnitude and direction of the resultant force may be found by a technique called *vector addition*. There are two graphical methods of performing vector addition: the *triangle of forces method* and the *parallelogram of forces method*.

Triangle of forces method

A simple procedure for the triangle of forces method of vector addition is as follows:

1. Draw a vector representing one of the forces, using an appropriate scale and in the direction of its line of action.

2. From the nose of this vector and using the same scale, draw a vector representing the second force in the direction of its line of action.

3. The resultant vector is represented in both magnitude and direction by the vector drawn from the tail of the first vector to the nose of the second vector.

Example 4.20

Determine the magnitude and direction of the resultant of a force of 15 N acting horizontally to the right and a force of 20 N, inclined at an angle of 60° to the 15 N force. Use the triangle of forces method.

Using the procedure given above and with reference to Figure 4.29:

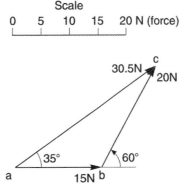

Figure 4.29

1. **ab** is drawn 15 units long horizontally

2. From b, **bc** is drawn 20 units long, inclined at an angle of 60° to ab. (Note, in angular measure, an angle of 60° from **ab** means 60° is in an anticlockwise direction.)

3. By measurement, the resultant ac is 30.5 units long inclined at an angle of 35° to **ab**. That is, the resultant force is **30.5 N** inclined at an angle of **35°** to the 15 N force.

Parallelogram of forces method

A simple procedure for the parallelogram of forces method of vector addition is as follows:

1. Draw a vector representing one of the forces, using an appropriate scale and in the direction of its line of action.

2. From the tail of this vector and using the same scale draw a vector representing the second force in the direction of its line of action.

3. Complete the parallelogram using the two vectors drawn in 1 and 2 as two sides of the parallelogram.

4. The resultant force is represented in both magnitude and direction by the vector corresponding to the diagonal of the parallelogram drawn from the tail of the vectors in 1 and 2.

Example 4.21

Use the parallelogram of forces method to find the magnitude and direction of the resultant of a force of 250 N acting at an angle of 135° and a force of 400 N acting at an angle of -120°.

Using the procedure given above and with reference to Figure 4.30:

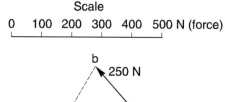

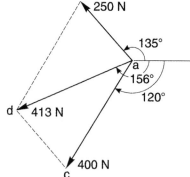

Figure 4.30

1. **ab** is drawn at an angle of 135° and 250 units in length

2. **ac** is drawn at an angle of -120° and 400 units in length

3. **bd** and **cd** are drawn to complete the parallelogram

4. **ad** is drawn. By measurement, **ad** is 413 units long at an angle of −156°. That is, the resultant force is **413 N** at an angle of **−156°**.

Test your knowledge 4.17

Find the magnitude and direction of the two forces given, using the triangle of forces method.

First force: 1.5 kN acting at an angle of 30°

Second force: 3.7 kN acting at an angle of −45°

Test your knowledge 4.18

Determine the magnitude and direction of the resultant of forces of 23.8 N at −50° and 14.4 N at 215° using the parallelogram of forces method.

Resultant of more than two coplanar forces

For the three coplanar forces F_1, F_2 and F_3 acting at a point as shown in Figure 4.31, the vector diagram is drawn using the nose-to-tail method. The procedure is:

1. Draw **0a** to scale to represent force F_1 in both magnitude and direction (see Figure 4.32).

2. From the nose of **0a**, draw **ab** to represent force F_2.

3. From the nose of **ab**, draw **bc** to represent force F_3.

4. The resultant vector is given by length **0c** in Figure 4.32. The direction of resultant **0c** is from where we started, i.e. point 0, to where we finished, i.e. point c. When acting by itself, the resultant force, given by **0c**, has the same effect on the point as forces F_1, F_2 and F_3 have when acting together. The resulting vector diagram of Figure 4.32 is called the *polygon of forces*.

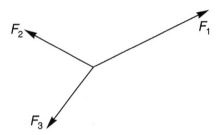

Figure 4.31 *Three forces acting on a point*

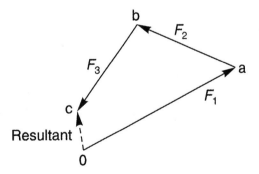

Figure 4.32 *Using the 'nose-to-tail' method to construct a parallelogram of forces*

Equilibrium

When three or more coplanar forces are acting at a point and the vector diagram closes, there is no resultant (in other words, the value of the resultant force is zero). The forces acting at the point are then said to be in *equilibrium*.

Example 4.22

Determine graphically the magnitude and direction of the resultant of these three coplanar forces, which may be considered as acting at a point. Force F_A, 12 N acting horizontally to the right; force F_B, 7N inclined at 60° to force F_A; force F_C, 15 N inclined at 150° to force F_A.

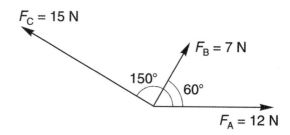

$F_C = 15$ N

$F_B = 7$ N

150°   60°

$F_A = 12$ N

Figure 4.33

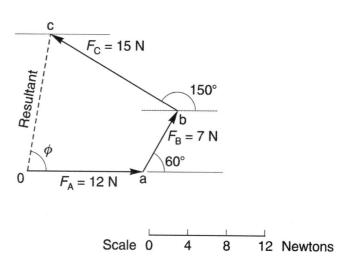

c

Resultant

$F_C = 15$ N

150°

b

$F_B = 7$ N

ϕ

60°

0   $F_A = 12$ N   a

Scale 0   4   8   12 Newtons

Figure 4.34

The space diagram is shown in Figure 4.33. The vector diagram (Figure 4.34) is produced as follows:

1.   **0a** represents the 12 N force in magnitude and direction

2.   From the nose of **0a**, **ab** is drawn inclined at 60° to **0a** and 7 units long.

3.   From the nose of **ab**, **bc** is drawn 15 units long inclined at 150° to **0a** (i.e. 150° to the horizontal).

4.   **0c** represents the resultant. By measurement, the resultant is **13.8 N** inclined at **80°** to the horizontal.

Hence the resultant of the three forces F_A, F_B and F_C is a force of **13.8 N** at **80°** to the horizontal.

Example 4.23

A load of 200 N is lifted by two ropes connected to the same point on the load, making angles of 40° and 35° with the vertical. Determine graphically the tensions in each rope when the system is in equilibrium.

The space diagram is shown in Figure 4.35.

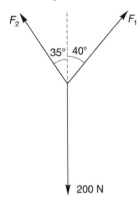

Figure 4.35

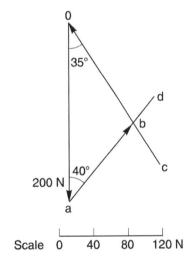

Scale   0   40   80   120 N

Figure 4.36

Test your knowledge 4.19

Find the magnitude and direction of the resultant of the following three coplanar forces:

First force: 23 kN acting at 80° to the horizontal

Second force: 30 kN acting at an angle of 37° to the first force

Third force: 15 kN acting at an angle of 70° to the second force

Since the system is in equilibrium, the vector diagram must close. The vector diagram (Figure 4.36) is drawn as follows:

1. The load of 200 N is drawn vertically as shown by **0a**.
2. The direction only of force F_1 is known, so from point a, **ad** is drawn 40° to the vertical.
3. The direction only of force F_2 is known, so from point o, **0c** is drawn at 35° to the vertical.
4. Lines **ad** and **0c** cross at point b. Hence the vector diagram is given by triangle 0ab. By measurement, **ab** is 119 N and 0b is 133 N.

Thus the tensions in the ropes are $F_1 = $ **119 N** and $F_2 = $ **133 N**.

Motion

Speed

Speed is the rate of covering distance and is given by:

$$\text{Speed} = \frac{\text{distance travelled}}{\text{time taken}}$$

The usual units for speed are metres per second (m/s or ms^{-1}) or kilometres per hour (km/h or km h^{-1}). Thus if a person walks 10 kilometres in 2 hours, the speed of the person is 10/2 or 5 kilometres per hour.

The symbol for the SI unit of speed (and velocity) can be written as m s^{-1} (using index notation) or simply as m/s.

Example 4.24

A man walks 600 metres in 5 minutes. Determine his speed expressed in (a) metres per second and (b) kilometres per hour.

(a) Here, the distance travelled = 600 m and the time taken = 5 minutes. We need to convert the time taken to seconds. Since there are 60 seconds in one minute we can simply multiply the time taken expressed in minutes by 60. Hence time taken = 5 × 60 = 300 seconds.

$$\text{Thus,} \quad \text{speed} = \frac{\text{distance travelled}}{\text{time taken}} = \frac{600\ \text{m}}{300\ \text{s}} = 2\text{m/s}$$

(b) Once again, the distance travelled = 600 m and the time taken = 5 minutes. In order to provide an answer in km/h we need to convert the distance expressed in metres to km and the time expressed in minutes to hours. To convert metres to km we divide by 1000 and to convert from minutes to hours we divide by 60. Hence the distance travelled = 600 / 1000 = 0.6 km and the time taken = 5 / 60 = 0.0833 hours.

$$\text{Thus,} \quad \text{speed} = \frac{\text{distance travelled}}{\text{time taken}} = \frac{0.6\ \text{km}}{0.0833\ \text{h}} = 7.2\ \text{km/h}$$

Note: To convert from m/s to km/h you can multiply by 3.6. Alternatively, to convert from km/h to m/s you can divide by 0.277.

Test your knowledge 4.20

A train is travelling at a constant speed of 25 metres per second for 16 kilometres. Find the time taken to cover this distance.

Test your knowledge 4.21

A car travels a distance of 120 km in 1 hour 15 minutes. Determine the average speed of the car (a) in km/h and (b) in m/s.

Distance/time graphs

A graph of distance travelled (on the vertical axis of the graph) against time (on the horizontal axis of the graph) is called a *distance/time graph*. Thus if an aircraft travels 500 km in its first hour of flight and 750 km in its second hour of flight, then after 2 hours, the total distance travelled is (500 + 750) km or 1250 km. The distance/time graph for this flight is shown in Figure 4.37.

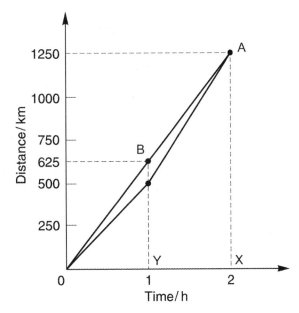

Figure 4.37 *A distance/time graph*

The average speed is given by:

$$\text{Average speed} = \frac{\text{total distance travelled}}{\text{total time taken}}$$

Thus, the average speed of the aircraft is:

$$\frac{(500 + 750)\ \text{km}}{(1 + 1)\ \text{h}} = \frac{1250}{2}\ \text{km/h} = 625\ \text{km/h}$$

If points 0 and A are joined in Figure 4.37, for any two points on line 0A the slope of line 0A is defined as:

$$\frac{\text{change in distance (vertical)}}{\text{change in time (horizontal)}}$$

For point A, the change in distance is AX (i.e. 1250 km) and the change in time is 0X (i.e. 2 hours). Hence the average speed is 1250/2 or 625 km/h.

Alternatively, for point B on line 0A, the change in distance is BY (i.e. 625 km) and the change in time is 0Y (i.e. 1 hour) hence the average speed is 625/1 or 625 km/h.

In general, the average speed of an object travelling between points M and N is given by the slope of line MN on the distance/time graph.

Speed/time graph

If a graph is plotted of speed against time, the area under the graph gives the distance travelled. This is demonstrated in Example 4.27.

Example 4.25

A person travels from point 0 to A, then from A to B and finally from B to C. The distances of A, B and C from 0 and the times, measured from the start to reach points A, B and C are as shown:

	A	B	C
Distance (m)	100	200	250
Time (s)	40	60	100

Plot the distance/time graph and determine the speed of travel for each of the three parts of the journey.

The vertical scale of the graph is distance travelled and the scale is selected to span 0 to 250 m, the total distance travelled from the start. The horizontal scale is time and spans 0 to 100 seconds, the total time taken to cover the whole journey. Coordinates corresponding to A, B and C are plotted and 0A, AB and BC are joined by straight lines. The resulting distance/time graph is shown in Figure 4.38.

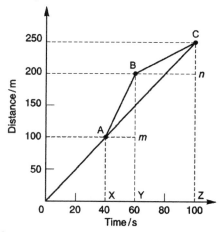

Figure 4.38

The speed is given by the slope of the distance/time graph. Speed for part 0A of the journey = slope of 0A = AX/ 0X

$$= \frac{100 \text{ m}}{40 \text{ s}} = \textbf{2.5 m/s}$$

Speed for part AB of the journey = slope of AB

$$= \frac{(200 - 100) \text{ m}}{(60 - 40) \text{ s}} = \frac{100 \text{ m}}{20 \text{ s}} = \textbf{5 m/s}$$

Speed for part BC of the journey = slope of BC

$$= \frac{(250 - 200) \text{ m}}{(100 - 60) \text{ s}} = \frac{50 \text{ m}}{40 \text{ s}} = \textbf{1.25 m/s}$$

Example 4.26

Determine the average speed (both in m/s and km/h) for the whole journey for the information given in Example 4.25.

$$\text{Average speed} = \frac{\text{total distance travelled}}{\text{total time taken}} = \text{slope of line 0C.}$$

From Figure 4.38,

$$\text{slope of line 0C} = \frac{250 \text{ m}}{100 \text{ s}} = \textbf{2.5 m/s}$$

Now $2.5 \text{ m/s} = \dfrac{2.5 \text{ m}}{1 \text{ s}} \times \dfrac{1 \text{ km}}{1000 \text{ m}} \times \dfrac{3600 \text{ s}}{1 \text{ h}} = 2.5 \times 3.6 \text{ km/h} = \textbf{9 km/h}$

Example 4.27

The motion of an object is described by the speed/time graph given in Figure 4.39. Determine the distance covered by the object when moving from 0 to B.

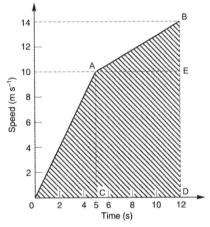

Figure 4.39

The distance travelled is given by the area beneath the speed/time graph, shown shaded in Figure 4.39.

Area of triangle 0AC = ½ x base x perpendicular height
= ½ x 5 s x 10 m/s = 25 ms/s = 25 m

The area of rectangle AEDC = base x height
= (12 – 5) s x (10 – 0) m/s = 70 ms/s = 70 m

Area of triangle ABE = ½ x base x perpendicular height
= ½ x (12 – 5) s x (14 – 10) m/s
= ½ x 7 s x 4 m/s = 14 ms/s = 14 m

Hence the distance covered by the object moving from 0 to B is
(25 + 70 + 14) m = **109 m**

Velocity

The velocity of an object is the speed of the object *in a specified direction*. Thus if an aircraft is flying due south at 500 kilometres per hour, its speed is 500 kilometres per hour, but its velocity is 500 kilometres per hour *due south*. It follows that if the aircraft had flown in a circular path for one hour at a speed of 500 kilometres per hour (so that one hour after taking off it is once again over the airport) its average velocity in the first hour of flight would be zero. It should be noted that, even though the aircraft has covered a distance whilst in flight, its *displacement* (i.e. the distance between its starting point and ending point) is zero. Hence, when considering the way in which something moves, the difference between speed and velocity is just as important as the difference between distance and displacement!

Average velocity is given by:

$$\text{Average velocity} = \frac{\text{distance travelled in a particular direction}}{\text{time taken}}$$

If an aircraft flies from place O to place A, a distance of 300 kilometres in one hour, A being due north of O, then OA in Figure 4.40 represents the first hour of flight. It then flies from A to B, a distance of 400 kilometres during the second hour of flight, B being due east of A, thus AB in Figure 4.40 represents its second hour of flight. Its average velocity for the two hour flight is:

$$\frac{\text{distance OB}}{\text{time taken}} = \frac{\text{distance OB}}{2 \text{ hours}} = \frac{500 \text{ km}}{2 \text{ h}} = 250 \text{ km/h}$$

Note that this velocity is in the direction OB.

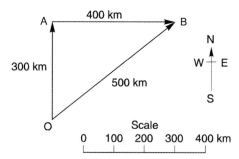

Figure 4.40

A graph of velocity (on the vertical axis of the graph) against time (on the horizontal axis of the graph) is called a *velocity/time graph*. The graph shown in Figure 4.41 represents an aircraft flying for 3 hours at a constant speed of 600 kilometres per hour in a specified direction. The shaded area represents velocity (vertically) multiplied by time (horizontally) and has units of:

i.e. kilometres and it represents the distance travelled in a specified direction.

A coach travels from town A to town B, a distance of 40 kilometres at an average speed of 60 kilometres per hour. It then travels from town B to town C, a distance of 50 kilometres in 25 minutes. Determine:
(a) the time taken to travel from A to B,
(b) the average speed of the coach from B to C,
(c) the average speed of the whole journey from A to C.

In this case, distance = 600 km/h × 3 h = 1800 km. Another method of determining the distance travelled is by using:

distance travelled = average velocity × time

Thus if a plane travels due south at 600 kilometres per hour for 20 minutes, the distance covered is:

h = 200 km

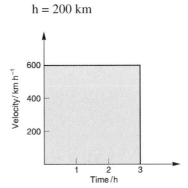

Figure 4.41

Acceleration

Acceleration is the rate of change of velocity with time. The average acceleration, a, is given by:

$$a = \frac{\text{change in velocity}}{\text{time taken}} = \frac{v - u}{t} \ \text{m/s}^2$$

where u is the initial velocity of an object in m/s, v is the final velocity in m/s and t is the time in seconds elapsing between the velocities of u and v.

Velocity/time graph

A graph of velocity (on the vertical axis) against time (on the horizontal axis) is called a velocity/time graph, as introduced above. From the velocity/time graph shown in Figure 4.42, the slope of line 0A is given by AX/0X. AX is the change in velocity from an initial velocity u of zero to a final velocity, v, of 4 metres per second. 0X is the time taken for this change in velocity, thus:

$$\frac{\text{AX}}{\text{0X}} = \frac{\text{change in velocity}}{\text{time taken}}$$

$$= \text{the acceleration in the first two seconds}$$

From the graph:

$$\frac{\text{AX}}{\text{0X}} = \frac{4 \ \text{m/s}}{2 \ \text{s}} = 2 \ \text{m/s}^2$$

i.e. the acceleration is 2 m/s^2. Similarly, the slope of line AB in Figure 4.40 is given by BY/AY, i.e. the acceleration between 2 s and 5 s is:

$$\frac{(8-4) \ \text{m/s}}{(5-2) \ \text{s}} = \frac{4 \ \text{m/s}}{3 \ \text{s}} = 1.33 \ \text{m/s}^2$$

As you can see, the slope of a line on a velocity/time graph gives the acceleration. Furthermore, like velocity, acceleration is a vector quantity and is more properly defined as rate of change of speed with respect to time *in a specified direction*.

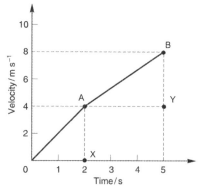

Figure 4.42 *Velocity/time graph*

Example 4.28

The speed of a car travelling along a straight road changes uniformly from zero to 50 km/h in 20 s. It then maintains this speed for 30 s and finally reduces speed uniformly to rest in 10 s. Draw the speed/time graph for this journey.

The vertical scale of the speed/time graph is speed (km h^{-1}) and the horizontal scale is time (s). Since the car is initially at rest, then at time 0 seconds, the speed is 0 km/h. After 20 s, the speed is 50 km/h, which corresponds to point A on the speed/time graph shown in Figure 4.43. Since the change in speed is uniform, a straight line is drawn joining points O and A. The speed is constant at 50 km/h for the next 30 s, hence, horizontal line AB is drawn in Figure 4.43 for the time period 20 s to 50 s. Finally, the speed falls from 50 km/h at 50 s to zero in 10 s, hence point C on the speed/time graph in Figure 4.43 corresponds to a speed of zero and a time of 60 s. Since the reduction in speed is uniform, a straight line is drawn joining BC. Thus, the speed/time graph for the journey is as shown in Figure 4.43.

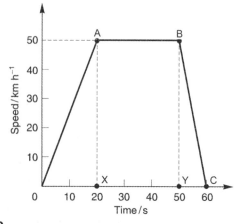

Figure 4.43

Example 4.29

For the speed/time graph shown in Figure 4.43, find the acceleration for each of the three stages of the journey.

From above, the slope of line 0A gives the uniform acceleration for the first 20 s of the journey

$$\text{Slope of 0A} = \frac{AX}{0X} = \frac{(50-0)\,km/h}{(20-0)\,s} = 50\frac{km}{h} \times \frac{1\,h}{20\,s}$$

Expressing 50 km/h in metre/second (m/s) units gives:

$$50\frac{km}{h} = \frac{50\,km}{1\,h} \times \frac{1000\,m}{1\,km} \times \frac{1\,h}{3600\,s} = \frac{50}{3.6}m/s$$

(Note: to change from km/h to m/s, divide by 3.6.)

Thus,

$$50\,km/h \times \frac{1}{20\,s} = \frac{50}{3.6}\,m/s \times \frac{1}{20\,s}$$

$$= 0.694\,m/s^2$$

i.e. the acceleration during the first 20 s is **0.694 m/s²**.

Acceleration is defined as:

$$\frac{\text{change in velocity}}{\text{time taken}} \text{ or } \frac{\text{change of speed}}{\text{time taken}}$$

as the car is travelling along a straight road. Since there is no change in speed for the next 30 s (line AB in Figure 4.43 is horizontal), then the acceleration for this period is zero. From above, the slope of line BC gives the uniform deceleration for the final 10 s of the journey.

$$\text{Slope of BC} = \frac{BY}{YC} = \frac{50\,km/h}{10\,s}$$

$$= \frac{50\,m}{3.6\,s} \times \frac{1}{10\,s}$$

$$= 1.39\,m/s^2$$

Hence the deceleration during the final 10 s is 1.39 m/s² and, since acceleration is the opposite of deceleration, we can conclude that the acceleration during the final 10 s is **−1.39 m/s²**.

Free-fall

When a dense object is falling under the influence of gravity it has a constant acceleration of approximately 9.81 m/s². We are thus able to determine the velocity of an object in *free-fall*. Note that low density objects (such as feathers) fall less quickly through the air. In a vacuum, however, all objects experience the same downward acceleration of 9.81 m/s².

Equations of motion

Earlier we defined acceleration as the change in velocity divided by the time. From this we concluded that:

$$\text{Acceleration,} \quad a = \frac{v-u}{t}$$

From which,

$$\frac{v-u}{t} = a$$

We can rearrange this formula to make velocity the subject by multiplying both sides by t, thus:

$$\frac{v-u}{t} \times t = a \times t$$

From which,

$$(v-u) \times \frac{t}{t} = a \times t$$

$$(v-u) \times 1 = a\,t$$

Adding u to each side gives,

$$v - u = a\,t$$

or

$$v - u + u = a\,t + u$$

Hence:

$$v = u + a\,t$$

where u is the initial velocity in m/s, v is the final velocity in m/s, a is the constant acceleration in m/s^2 and t is the time in s.

When a has a negative value it is more commonly referred to as *deceleration* or *retardation*. The equation $v = u + at$ is an important *equation of motion*.

Activity 4.5

A satellite is launched from a space vehicle by firing a thruster motor. During tests the following data was obtained during the first 10 s after separation:

Time (s)	0	2	4	6	8	10
Velocity (m/s)	4.5	5.2	5.9	6.6	7.3	8.1

Plot the velocity/time graph and use it to determine:

(a) the initial velocity of the satellite
(b) the acceleration produced by the thruster
(c) the average velocity during the 10 s period
(d) the velocity 5 seconds after firing the thruster
(e) the separation distance after the 10 s period.

Present your answer as a fully worked solution.

Example 4.30

A dense object is dropped from a tall building. Determine (a) its velocity after 2 s and (b) the increase in velocity during the third second, in the absence of all forces except that due to gravity.

The object is free-falling and thus has an acceleration, a, of approximately 9.81 m/s^2 (taking downward motion as positive).

We can determine the final velocity from:

$$v = u + a\,t$$

(a) The initial downward velocity, u, of the object is zero. The acceleration, a, is 9.81 m/s^2 downwards and the time during which the stone is accelerating is 2 s. Hence, the final final velocity is given by:

$$v = 0 + 9.81 \times 2 = 19.62 \text{ m/s}$$

i.e. the velocity of the stone after 2 s is approximately **19.62 m/s**.

(b) From part (a), the velocity after two seconds, u, is 19.62 m/s. The velocity after 3 s, applying $v = u + a\,t$ once again is:

$$v = 19.62 + 9.81 \times 3 = 49.05 \text{ m/s}.$$

Thus, the change in velocity during the third second is: (49.05 − 19.62) m/s, that is **29.43 m/s**.

Test your knowledge 4.23

1 A ship changes velocity from 15 km/h to 20 km/h in 25 min. Determine the average acceleration in m/s^2 of the ship during this time.

2 Determine how long it takes an object, which is free-falling, to change its speed from 100 km/h to 150 km/h, assuming all other forces, except that due to gravity, are neglected.

3 A car travelling at 50 km/h applies its brakes for 6 s and decelerates uniformly at 0.5 m/s. Determine its velocity in km/h after the 6 s braking period.

Example 4.31

A train travelling at 30 km/h accelerates uniformly to 50 km/h in 2 minutes. Determine the acceleration.

$$30 \text{ km}/\text{h} = \frac{30}{3.6} \text{m}/\text{s} = 8.3 \text{ m}/\text{s} \quad (\text{see Example 4.29})$$

$$50 \text{ km}/\text{h} = \frac{50}{3.6} \text{m}/\text{s} = 13.9 \text{ m}/\text{s}$$

$$2 \text{ min} = 2 \times 60 = 120 \text{ s}$$

Hence $u = 8.3$ m/s, $v = 13.9$ m/s and $t = 120$ s. Substituting these values in $v = u + at$ gives:

$$13.9 = 8.3 + (a \times 120)$$

$$13.9 - 8.3 = a \times 120$$

$$5.6 = a \times 120$$

$$a = \frac{5.6}{120} = 0.046 \text{ m}/\text{s}^2$$

Thus the uniform acceleration of the train is **0.046 m/s^2**.

Force, mass and acceleration

When an object is pushed or pulled, a force is applied to the object. This force is measured in newtons (N). The effects of pushing or pulling an object are:

(i) to cause a change in the motion of the object, and
(ii) to cause a change in the shape of the object.

If a change in the motion of the object, that is, its velocity changes from u to v, then the object accelerates. Thus, it follows that acceleration results from a force being applied to an object. If a force is applied to an object and it does not move, then the object changes shape, that is, deformation of the object takes place. Usually the change in shape is so small that it cannot be detected by just watching the object. However, when very sensitive measuring instruments are used, very small changes in dimensions can be detected.

A force of attraction exists between all objects. The factors governing the size of this force F are the masses of the objects and the distances between their centres.

Thus, if a person is taken as one object and the earth as a second object, a force of attraction exists between the person and the earth. This force is called the gravitational force and is the force which gives a person a certain weight when standing on the earth's surface. It is also this force which gives freely falling objects a constant acceleration in the absence of other forces.

Newton's laws of motion

To make a stationary object move or to change the direction in which the object is moving requires a force to be applied externally to the object. This concept is known as Newton's *first law of motion* and may be stated as:

> *An object remains in a state of rest, or continues in a state of uniform motion in a straight line, unless it is acted on by an externally applied force.*

Since a force is necessary to produce a change of motion, an object must have some resistance to a change in its motion. The force necessary to give a stationary pram a given acceleration is far less than the force necessary to give a stationary car the same acceleration. The resistance to a change in motion is called the *inertia* of an object and the amount of inertia depends on the mass of the object. Since a car has a much larger mass than a cycle, the inertia of a car is much larger than that of a cycle. Newton's *second law of motion* may be stated as:

> *The acceleration of an object acted upon by an external force is proportional to the force and is in the same direction as the force.*

Thus, force is proportional to acceleration, or force = a constant × acceleration, this constant of proportionality being the mass of the object, i.e. force = mass × acceleration

The unit of force is the newton (N) and is defined in terms of mass and acceleration. One newton is the force required to give a mass of 1 kilogram an acceleration of 1 metre per second squared.

Thus

$$F = m\,a$$

where F is the force in newtons (N), m is the mass in kilograms (kg) and a is the acceleration in metres per second squared (m/s²), i.e.

$$1\,\text{N} = \frac{1\ \text{kg m}}{\text{s}^2}$$

It follows that 1 m/s² = 1 N/kg. Hence a gravitational acceleration of 9.81 m/s² is the same as a gravitational field of 9.81 N/kg.

Newton's third law of motion may be stated as:

For every force, there is an equal and opposite reacting force.

Thus, an object on, say, a table, exerts a downward force on the table and the table exerts an equal upward force on the object, known as a reaction force or just a reaction.

Example 4.32

Calculate the force needed to accelerate a boat of mass 20 tonne uniformly from rest to a speed of 21.6 km/h in 10 minutes.

The mass of the boat, m, is 20 t, that is 20000 kg. The equation of motion, $v = u + a\,t$, can be used to determine the acceleration a. The initial velocity, u, is zero. The final velocity, v is 21.6 km/h, that is, 21.6/3.6 or 6 m/s. The time, t, is 10 min (or 600 s).

Thus 6 = 0 + a x 600   or $a = \dfrac{6}{600} = \textbf{0.01 m/s}^2$

From Newton's second law, $F = m\,a$, i.e.

Force = 20000 x 0.01 N = **200 N**

Example 4.33

The moving head of a machine tool requires a force of 1.2 N to bring it to rest in 0.8 s from a cutting speed of 30 m/min. Find the mass of the moving head.

From Newton's second law, $F = m\,a$, thus $m = F / a$, where force is given as 1.2 N. The equation of motion $v = u + a\,t$ can be used to find acceleration a, where $v = 0$, $u = 30$ m/min, that is 30/60 or 0.5 m/s, and $t = 0.8$ s. Thus,

0 = 0.5 + a x 0.8, i.e. $a = \dfrac{0.5}{0.8} = -0.625\,\text{m/s}^2$

i.e. a retardation of 0.625 m/s²

Thus the mass, m = 1.2 / 0.625 = **1.92 kg**

Example 4.34

Find the weight of an object of mass 1.6 kg at a point on the earth's surface, where the gravitational field is 9.81 N/kg.

The weight of an object is the force acting vertically downwards due to the force of gravity acting on the object. Thus:

Weight = force acting vertically downwards

= mass × gravitational field

= 1.6 × 9.81 = **15.696 N**

Example 4.35

A bucket of cement of mass 40 kg is tied to the end of a rope connected to a hoist. Calculate the tension in the rope when the bucket is suspended but stationary. Take the gravitational field, g, as 9.81 N/kg.

The tension in the rope is the same as the force acting in the rope. The force acting vertically downwards due to the weight of the bucket must be equal to the force acting upwards in the rope, i.e. the tension, T.

Weight of bucket of cement, $F = mg = 40 \times 9.81 = 392.4$ N.

Thus, the tension in the rope is also **392.4 N**.

Example 4.36

The bucket of cement in Example 4.35 is now hoisted vertically upwards with a uniform acceleration of 0.4 m/s². Calculate the tension in the rope during the period of acceleration.

With reference to Figure 4.44(a), the forces acting on the bucket are:
(i)   a tension (or force) of T acting in the rope;
(ii)   a force of mg acting vertically downwards, i.e. the weight of the bucket and cement

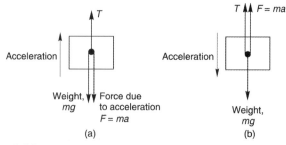

Figure 4.44

The resultant force $F = T - mg$. Hence $ma = T - mg$.

$$40 \times 0.4 = T - 40 \times 9.81 \text{ giving } T = 408.4 \text{ N}$$

By comparing this result with that of Example 4.35, it can be seen that there is an increase in the tension in the rope when an object is accelerating upwards.

Test your knowledge 4.24

1   A lorry of mass 1350 kg accelerates uniformly from 9 km/h to reach a velocity of 45 km/h in 18 s. Determine (a) the acceleration of the lorry, and (b) the uniform force needed to accelerate the lorry.

2   The tension in a rope lifting a crate vertically upwards is 2.8 kN. Determine its acceleration if the mass of the crate is 270 kg.

Work

If a body moves as a result of a force being applied to it, the force is said to do work on the body. The amount of work done is the product of the applied force and the distance, i.e.

work done = force × distance moved in the direction of the force

The unit of work is the *joule*, J, which is defined as the amount of work done when a force of 1 N acts for a distance of 1 m in the direction of the force. Thus,

$$1 \text{ J} = 1 \text{ N m}$$

If a graph is plotted of experimental values of force (on the vertical axis) against distance moved (on the horizontal axis) a force/distance graph or *work diagram* is produced. The area under the graph represents the work done. For example, a constant force of 20 N used to raise a load a height of 8 m may be represented on a force/distance graph as shown in Figure 4.45(a). The area under the graph, shown shaded, represents the work done. Hence:

work done = 20 N × 8 m = 160 J

Similarly, a spring extended by 20 mm by a force of 500 N may be represented by the work diagram shown in Figure 4.45(b).

work done = shaded area = ½ × base × height

$$= ½ × (20 × 10^{-3}) × 500 \text{ N} = 5 \text{ J}$$

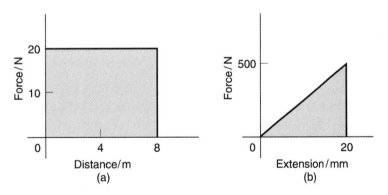

Figure 4.45 *Work diagrams*

Example 4.37

Calculate the work done when a force of 40 N pushes an object a distance of 500 m in the same direction as the force.

work done = force distance moved in the direction of the force

40 N 500 m = 20,000 J (since 1 J = 1 Nm)

Thus work done = **20 kJ**

Example 4.38

A motor supplies a constant force of 1 kN which is used to move a load a distance of 5 m. The force is then changed to a constant 500 N and the load is moved a further 15 m. Draw the force/distance graph for the operation and from the graph determine the work done by the motor.

The force/distance graph of work diagram is shown in Figure 4.46. Between points A and B a constant force of 1000 N moves the load 5 m; between points C and D a constant force of 500 N moves the load from 5 m to 20 m.

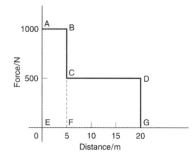

Figure 4.46

Total work done = area under the force/distance graph

$$= \text{area ABFE} + \text{area CDGF}$$

$$= (1000 \text{ N} \times 5 \text{ m}) + (500 \text{ N} \times 15 \text{ m})$$

$$= 5000 \text{ J} + 7500 \text{ J} = 12{,}500 \text{ J} = \textbf{12.5 kJ}.$$

Example 4.39

A spring, initially in a relaxed state, is extended by 100 mm. Determine the work done by using a work diagram if the spring requires a force of 0.6 N per mm of stretch.

Force required for a 100 mm extension = 100 mm × 0.6 N / mm = 60 N.

Figure 4.47 shows the force/extension graph or work diagram representing the increase in extension in proportion to the force, as the force is increased from 0 to 60 N. The work done is the area under the graph (shown shaded).

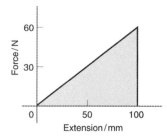

Figure 4.47

Hence,

$$\text{work done} = \tfrac{1}{2} \times \text{base} \times \text{height} = \tfrac{1}{2} \times 100 \text{ mm} \times 60 \text{ N}$$

$$= \tfrac{1}{2} \times 100 \times 10^{-3} \text{ mm} \times 60 \text{ N} = \tfrac{1}{2} \times 0.1 \text{ m} \times 60 \text{ N} = \textbf{3 J}$$

Energy

Forms of energy

Energy is the capacity, or ability, to do work. The unit of energy is the joule (J) the same as for work. Energy is expended when work is done. There are several forms of energy and these include:

- mechanical energy
- heat or thermal energy
- electrical energy
- chemical energy
- nuclear energy
- light energy
- sound energy.

Energy may be converted from one form to another. The *principle of conservation of energy* states that the total amount of energy remains the same in such conversions, i.e. energy cannot be created or destroyed. Some examples of energy conversions include:

- Mechanical energy is converted to electrical energy by a generator.
- Electrical energy is converted to mechanical energy by a motor.
- Heat energy is converted to mechanical energy by a steam engine.
- Mechanical energy is converted to heat energy by friction.
- Heat energy is converted to electrical energy by a solar cell.
- Electrical energy is converted to heat energy by an electric fire.
- Heat energy is converted to chemical energy by living plants.
- Chemical energy is converted to heat energy by burning fuels.
- Heat energy is converted to electrical energy by a thermocouple.
- Chemical energy is converted to electrical energy by batteries.
- Electrical energy is converted to light energy by a light bulb.
- Sound energy is converted to electrical energy by a microphone.
- Electrical energy is converted to chemical energy by electrolysis.

Efficiency

Efficiency is defined as the ratio of the useful output energy to the input energy. The symbol for efficiency is η (or 'eta'). Hence:

Efficiency has no units and is often stated as a percentage. A perfect machine would have an efficiency of 100%. However, all machines have an efficiency lower than this due to friction and other losses. Thus, if the input energy to a motor is 1000 J and the output energy is 800 J then the efficiency is:

$$\text{efficiency, } \eta = \frac{80}{100} \times 100\%$$

Hence efficiency = 80%

Example 4.40

A machine exerts a force of 200 N in lifting a mass through a height of 6 m. If 2 kJ of energy are supplied to it, what is the efficiency of the machine?

Work done in lifting mass = force × distance moved

= weight of body × distance moved

= 200 N × 6 m = 1200 J

= useful energy output

Now, energy supplied = 2 kJ = 2000 J

and efficiency, $\eta = \dfrac{\text{useful output energy}}{\text{input energy}}$

Thus efficiency, $\eta = \dfrac{1200}{2000} = 0.6$ = **60%**

Example 4.41

4 kJ of energy are supplied to a machine used for lifting a mass. The force required is 800 N. If the machine has an efficiency of 50%, to what height will it lift the mass?

efficiency, $\eta = \dfrac{\text{useful output energy}}{\text{input energy}}$

Thus efficiency, $\eta = 50\% = 0.5 = \dfrac{\text{output energy}}{4000 \text{ J}}$

From which

output energy = 0.5 × 4000 = 2000 J

Finally work done = force × distance moved

hence 2000 J = 800 N × height

from which, height = $\dfrac{2000 \text{ J}}{800 \text{ J}}$ = **2.5 m**

Example 4.42

A hoist exerts a force of 500 N in raising a load through a height of 20 m. The efficiency of the hoist gears is 75% and the efficiency of the motor is 80%. Calculate the input energy to the hoist.

The hoist system is shown diagrammatically in Figure 4.48.

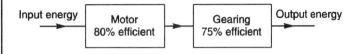

Figure 4.48

Output energy = work done = force × distance

$$= 500 \text{ N} \times 20 \text{ m} = 10\ 000 \text{ J}$$

For the gearing,

$$\text{efficiency, } \eta = \frac{\text{useful output energy}}{\text{input energy}}$$

Hence,

$$\text{efficiency, } \eta = 75\% = 0.75 = \frac{10000 \text{ J}}{\text{input energy}}$$

from which, the input energy to the gears = 10 000 / 0.75 = 13 333 J.

The input energy to the gears is the same as the output energy of the motor. Thus, for the motor,

$$\text{efficiency, } \eta = \frac{\text{useful output energy}}{\text{input energy}}$$

Hence,

$$\text{efficiency, } \eta = 80\% = 0.8 = \frac{13333 \text{ J}}{\text{input energy}}$$

Finally,

$$\text{input energy} = \frac{13333 \text{ J}}{0.8} = 16\ 670 \text{ J} = \textbf{16.67 kJ}.$$

Potential energy

Potential energy (*PE*) is the energy that a mass or body has by *virtue of its position*. In other words if a mass was to fall from a height it would release energy at the end of its fall. Potential energy is equal to the product of mass, gravitational force and height. Thus:

$$PE = m\,g\,h$$

PE will be in joules (J) when the mass, *m*, is expressed in kg, the gravitational force, *g*, is expressed in m/s^2, and the height, *h*, is expressed in m.

Kinetic energy

Kinetic energy (*KE*) is the energy that a mass or body has by *virtue of its motion*. In other words if a moving mass was to suddenly stop it would release energy when being brought to rest. Kinetic energy is equal to the product of half the mass and the square of the velocity. Thus:

$KE = \frac{1}{2} m v^2$

KE will be in joules (J) when the mass, *m*, is expressed in kg, and the velocity, *v*, is expressed in m/s .

Example 4.43

Determine the potential energy of a mass of 4 kg at a height of 20 m.

Now *PE* = *m g h* = 4 × 9.81 × 20 = **784.8 J**

Example 4.44

Determine the potential energy of a rocket travelling at 250 m/s if it has a mass of 41 kg.

Now *KE* = ½ *m v²* = 0.5 × 41 × 250²

= 0.5 × 41 × 62500

= 1 281 250 = 1 281.25 kJ = **1.28125 MJ**

Activity 4.6

Investigate the energy sources that are used in your home. Identify each source and explain how it arrives in your home. Explain what the energy is used for and how it is converted in other forms of energy. Also identify energy that is wasted and describe methods that are used to improve the efficiency of the energy conversion that takes place.

Present your answer in the form of a word processed report and include simple block diagrams that show the conversions that take place (see Figure 4.48).

Activity 4.7

Supplies of energy are rarely constant (consider, for example, the case of a wind generator). Describe a method of energy storage that can be applied to EACH of the following situations:

* a hydroelectric power station
* a rotating shaft
* a portable CD player.

Illustrate your answer with relevant sketches and explain the principle on which each of the methods operate. Present your answer in the form of a brief class presentation using appropriate visual aids.

Power

Power is a measure of the rate at which work is done or at which energy is converted from one form to another. Thus:

$$\text{Power, } P = \frac{\text{energy used}}{\text{time taken}}$$

or

$$\text{Power, } P = \frac{\text{work done}}{\text{time taken}}$$

The unit of power is the watt, W, where 1 W is equal to 1 joule per second. The watt is a small unit for many purposes and a larger unit called the kilowatt, kW, is used, where 1 kW = 1000 W. The power output of a motor which does 120 kJ of work in 30 s is thus given by:

$$\text{Power, } P = \frac{120 \text{ kJ}}{30 \text{ s}} = 4 \text{ kW}$$

Since work done = force × distance, then:

$$\text{Power, } P = \frac{\text{work done}}{\text{time taken}} = \frac{\text{force} \times \text{distance}}{\text{time taken}}$$

$$= \text{force} \times \frac{\text{distance}}{\text{time taken}}$$

However,

$$\frac{\text{distance}}{\text{time taken}} = \text{velocity}$$

Hence,

$$\text{power} = \text{force} \times \text{velocity}$$

$$P = Fv$$

In electrical circuits,

$$\text{power} = \text{current} \times \text{potential difference}$$

$$P = IV$$

Example 4.45

The output power of a motor is 8 kW. How much work does it do in 30 s?

Now,

$$\text{power} = \frac{\text{work done}}{\text{time taken}}$$

from which, work done = power × time

$$= 8000 \text{ W} \times 30 \text{ s}$$

$$= 240\,000 \text{ J} = \textbf{24 kJ}$$

Example 4.46

Calculate the power required to lift a mass through a height of 10 m in 20 s if the force required is 3924 N.

Work done = force × distance moved = 3924 N × 10 m = 39 240 J

Now, $\text{power} = \dfrac{\text{work done}}{\text{time taken}} = \dfrac{39\,240 \text{ J}}{20 \text{ s}} = 1962 \text{ W} = \textbf{1.962 kW}$

Example 4.47

A 220 V motor is supplied with a current of 6 A. How much power is supplied to the motor?

Now power = current × voltage = 6 A × 220 V = 1320 = **1.32 kW**

Example 4.48

If the motor in Example 4.47 is 60% efficient determine the mechanical power produced by the motor and the vertical distance that it will lift a mass of 10 kg in 6 s.

Here, efficiency, $\eta = \dfrac{\text{output power}}{\text{input power}}$

Hence,

output power = efficiency × input power

$$= 0.6 \times 1.32 \text{ kW} = \textbf{782 W}$$

Now, work done = force × distance

But the force required = $m\,g$ = 10 kg × 9.81 m/s² = 981 N

Thus, work done = 981 N × distance

Since, work done = power × time taken

981 N × distance = power × time taken

Hence, $\text{distance} = \dfrac{\text{power} \times \text{time taken}}{981 \text{ N}} = \dfrac{782 \text{ W} \times 6 \text{ s}}{981 \text{ N}} = \textbf{4.78 m}$

Test your knowledge 4.26

1   10 kJ of work is done by a force in moving a body uniformly through 125 m in 50 s. determine (a) the value of the force and (b) the power.

2   A planing machine has a cutting stroke of 2 m and the stroke takes 4 s. If the constant resistance to the cutting tool is 900 N calculate for each cutting stroke (a) the power consumed at the tool point and (b) the power input to the system if the system has an efficiency of 75%.

3   An electric motor provides power to a winding machine. The input power to the motor is 2.5 kW and the overall efficiency is 60%. Calculate (a) the output power of the machine and (b) the rate at which it can raise a 300 kg load vertically upwards.

Review questions

1   A charge of 1.8 C is transferred during a time interval of 36 s. What current will flow?

2   State Ohm's law.

3   A current of 40 mA flows in a resistor of 2.5 kΩ. What voltage will be dropped across the resistor?

4   Resistors of 27 Ω and 63 Ω are connected in series. Determine the current that will flow if the series combination is connected to an 18 V battery. Also determine the voltage dropped across each resistor.

5   Resistors of 10 Ω and 20 Ω are connected in parallel. Determine the current that will flow if the parallel combination is connected to an 6 V battery. Also determine the current flowing in each resistor.

6   A copper coil has a resistance of 500 Ω at 0°C. If copper has a temperature coefficient of resistance of 0.0043/°C determine the resistance of the coil at (a) 20°C and (b) 100°C.

7   State THREE applications that make use of the magnetic effect of an electric current.

8   State Faraday's laws of electromagnetic induction.

9   A copper conductor has a length of 100 mm. Determine the e.m.f. induced in the conductor if the conductor is moved at a constant velocity of 4.5 m/s at right angles to a uniform magnetic field of 1.1 T.

10   A sine wave voltage has an amplitude of 20 V and a frequency of 100 Hz. Sketch a waveform showing two complete cycles of the waveform and label the axes of voltage and time.

11   Sketch a diagram to illustrate the principle of the AC generator. Label your diagram.

12   Find the magnitude and direction of the result of the following coplanar forces: 12 kN acting at an angle of 30° to the horizontal and 20 kN acting at an angle of –45° to the horizontal.

13   Explain what is meant by the term *equilibrium* when applied to a system of forces.

14   A conveyor belt travels at a constant velocity of 0.5 m/s. How long will it take for an object travelling on the belt to move through a distance of 25 m?

15   A projectile travels at a constant velocity of 150 m/s. How far will the projectile travel in a time interval of 4.5 s?

16   A concrete block is dropped from the centre of a suspension bridge that spans a gorge. Determine the time taken for the block to reach the bottom of the gorge if the height at the centre of the bridge is 100 m and the block starts with an initial velocity of zero.

17   The speed of a car travelling along a straight road changes at a constant rate from 0 to 45 km/h in 15 s. The car then maintains a constant speed of 45 km/h for 25 s. Finally the car decelerates at a constant rate to become stationary again during a further time period of 10 s. Sketch the speed/time graph for the journey. Label your graph and also calculate the initial acceleration, final deceleration, and total distance travelled by the car.

18   A goods train accelerates from 50 km/h to 80 km/h in a time interval of 45 s. Determine the acceleration produced. Also Determine how long it would take to bring the train to rest if it is travelling at 80 km/h and the braking system provides a constant deceleration of 2 m/s^2.

19   A force of 1.5 kN is exerted on a mass of 1.2 kg. What acceleration will be produced?

20   State Newton's first and second laws of motion.

21   Find the weight of an object having a mass of 25 kg at a point on the surface of a planet where the gravitational field is 3.924 N/kg.

22   The tension in a rope lifting a load vertically is 3.6 kN. Determine the acceleration of the load if the mass of the crate is 347 kg.

23   Determine the potential energy of a 15 kg mass when it is 20 m above the surface of the earth.

24   Determine the kinetic energy of a vehicle having a mass of 1.5 tonne travelling at a speed of 60 km/h.

25   Calculate the work done when a force of 120 N pushes an object through a distance of 25 m in the same direction as the applied force.

26   Define the term 'power' in a mechanical system.

27   8 kJ of work is done by a force in moving a body uniformly through a distance of 8.5 m in a time interval of 34 s. Determine the force applied and the power supplied.

28   Describe THREE forms of energy and explain how these forms of energy can be converted from one form to another form.

29   With reference to a mechanical system, define the term 'efficiency'.

30   A 220 V motor consumes 4 A. If the motor is 60% efficient, determine the mechanical power supplied by the motor.

Useful formulae

Area, length and volume

Shape	Feature	Formula
Rectangle	Area Perimeter	$A = l \times b$
Parallelogram	Area	$A = b \times h$
Triangle	Area	$A = \frac{1}{2} \times b \times h$ $A = \frac{1}{2}\, bc \sin A$ $A = \sqrt{s(s-a)(s-b)(s-c)}$ Where $\quad s = \dfrac{a+b+c}{2}$
Trapezium	Area	$A = \frac{1}{2} \times h \times (a + b)$
Circle	Area Circumference	$A = \pi r^2 = \pi d^2/4$ $L = \pi d = 2\pi r$
Semicircle	Area Circumference	$A = \pi r^2/2 = \pi d^2/8$ $L = \frac{1}{2}\pi d = \pi r$
Sector of a circle	Area	$\theta/360 \times \pi r^2$ (θ in degrees) or $\frac{1}{2}r^2\,\theta$ (θ in radians)
Cone	Surface area Volume	$A = 2\pi r h + 2\pi r^2 = 2\pi(h + r)$ $V = 1/3\,\pi r^2 h$
Frustum of a cone	Curved area Total surface area Volume	$\pi l(R + r)$ $\pi l(R + r) + \pi R^2 + \pi r^2$ $1/3\,\pi h(R^2 + Rr + r^2)$
Sphere	Surface area Volume	$A = 4\pi r^2$ $V = 4/3\pi r^3$
Hemisphere	Surface area Volume	$A = 2\pi r^2$ $V = 2/3\pi r^3$
Pyramid	Surface area Volume	Sum of areas of the triangles forming the sides plus the area of the base $V = 1/3Ah$
Prism	Surface area Volume	Sum of longitudinal area plus area of the two ends C.s.a. × length

Density

$$\text{Density} = \frac{\text{mass}}{\text{volume}}$$

Motion

$$\text{Speed} = \frac{\text{distance}}{\text{time}}$$

$$\text{Acceleration,} \ a = \frac{v - u}{t}$$

Final velocity, $v = u + a\,t$

Energy

Potential energy, $PE = m\,g\,h$

Kinetic energy, $KE = \frac{1}{2}\,m\,v^2$

Sine of an angle

$$\text{Sine of an angle} = \frac{\text{opposite}}{\text{hypotenuse}}$$

$$\sin \theta = \frac{\text{opposite}}{\text{hypotenuse}}$$

Pythagoras' rule

$$(\text{hypotenuse})^2 = (\text{opposite})^2 + (\text{adjacent})^2$$

Electric charge

Charge = current $\times$ time

Ohm's Law

$$V = I \times R \quad \text{or} \quad I = \frac{V}{R} \quad \text{or} \quad R = \frac{V}{I}$$

Electrical power

$$P = I \times V \quad \text{or} \quad I = \frac{P}{V} \quad \text{or} \quad V = \frac{P}{I}$$

Resistors in series

$$R = R_1 + R_2$$

Resistors in parallel

$$\frac{1}{R} = \frac{1}{R_1} + \frac{1}{R_2} \quad \text{or} \quad R = \frac{R_1 \times R_2}{R_1 + R_2}$$

AC waveforms

$$f = 1/t \ \text{ or } \ t = 1/f$$

$$\text{amplitude} = V_m$$

$$\text{peak-peak value} = 2 \times V_m$$

Induced e.m.f.

$$E = B \, l \, v$$

$$E = B \, l \, v \sin \theta$$

Mechanics

$$\text{Force} = \text{mass} \times \text{acceleration}$$

$$F = m \, a$$

$$\text{weight} = m \, g$$

$$\text{work done} = \text{force} \times \text{distance}$$

$$\text{power} = \frac{\text{work done}}{\text{time taken}}$$

Efficiency

$$\text{efficiency} = \frac{\text{output energy}}{\text{input energy}}$$

$$\text{efficiency} = \frac{\text{output}}{\text{input}} \times 100 \, \%$$

Answers to test your knowledge questions

Chapter 4

4.1 (a) 9.871×10^3
(b) 3.34×10^{-1}
(c) 1.4576×10^5
(d) 5×10^{-4}

4.2 (a) $6 \times 10^{-2} \, \text{m}^2$
(b) $6 \times 10^2 \, \text{m}^2$
(c) $6 \times 10^4 \, \text{m}^2$

4.3 (a) $1.6 \times 10^3 \, \text{cm}^3$
(b) $1.6 \times 10^{-3} \, \text{m}^3$
(c) $1.6 \times 10^6 \, \text{mm}^3$

4.4 1 $6.25 \times 10^3 \, \text{kg/m}^3$
2 50 litres
3 $4 \times 10^3 \, \text{kg/m}^3$

4.5 1 150 C
2 8 s
3 6 mA $(6 \times 10^{-3} \, \text{A})$

4.6 1 4 kΩ, 1 mA
2 (a) 48 Ω
(b) 20 kΩ

4.7 0.25 A, 4 V

4.8 (a) nickel, copper, aluminium
(b) carbon
(c) constantan
(d) nickel

4.9 (a) 329.12 Ω
(b) 35 Ω
(c) 5,150 Ω (5.15 kΩ)

4.10 (a) P
(b) T
(c) S

	(d)	Q
	(e)	P, Q, R
	(f)	S

4.13 5.76 V, 1.6 A

4.14 0.16 V

4.15 (a) AC
     (b) 1.2 A
     (c) 2.4 A
     (d) 0.016 s (16 ms)
     (e) 62.5 Hz

4.17 4.3 kN at $-25°$

4.18 26.7 N at $-82°$

4.19 53.5 kN at $37°$ to the first force ($117°$ to the horizontal)

4.20 10 min 45 sec

4.21 (a) 96 km/h
     (b) 26.667 m/s

4.22 (a) 40 min
     (b) 120 km/h
     (c) 1.385 m/s

4.23 1 9.26×10^{-4} m/s
     2 1.42 s
     3 39.2 km/h

4.24 1 (a) 0.555 m/s^2
       (b) 750 N
     2 0.560 m/s^2

4.25 1 490 J
     2 (a) 2.5 J
       (b) 2.1 J
     3 14.72 kJ

4.26 1 (a) 80 N
       (b) 200 W
     2 (a) 450 W
       (b) 600 W
     3 (a) 1500 W (1.5 kW)
       (b) 510 mm/s

Answers to review questions

Chapter 4

1   50 mA

3   100 V

4   0.2 A, 5.4 V, 12.6 V

5   0.9 A, 0.6 A, 0.3 A

6   (a) 543 Ω
        (b) 715 Ω

9   0.495 V

12   25.85 kN at $-18.4°$ to the horizontal

14   50 s

15   675 m

16   4.52 s

17   3 m/s^2, 4.5 m/s^2, 1687.5 m (1.6875 km)

18   0.667 m/s^2, 40 s

19   1.25 m/s^2

21   98.1 N

22   0.565 m/s^2

23   2943 J (2.943 kJ)

24   12.5 kJ

25   3 kJ

27   0.941 kN, 235.3 Ω

30   528 W

Index